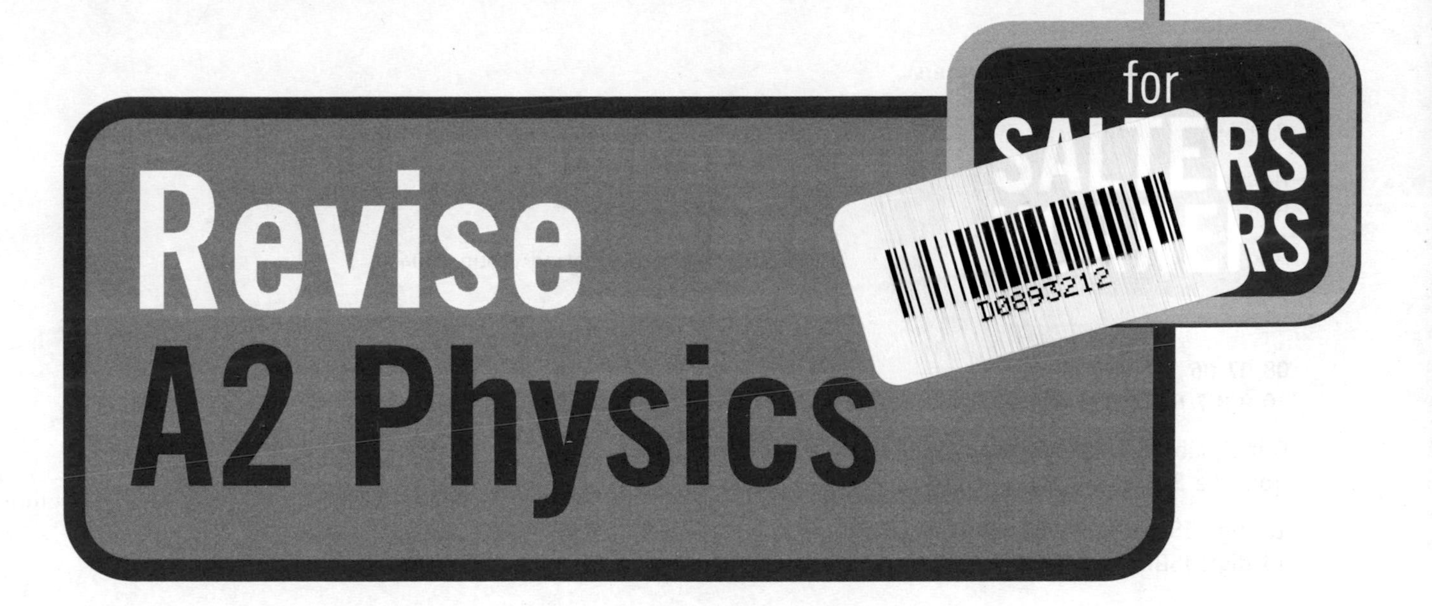

Richard Laird
Charlie Milward
Elizabeth Swinbank

Heinemann Educational Publishers
Halley Court, Jordan Hill, Oxford OX2 8EJ
Part of Harcourt Education

Heinemann is the registered trademark of
Harcourt Education Limited

First published 2004

08 07 06
10 9 8 7 6 5 4 3

British Library Cataloguing in Publication Data is available
from the British Library on request.

10-digit ISBN: 0 435582 08 9
13-digit ISBN: 978 0 435582 08 1

Edited by Anne Trevillion
Index compiled by Paul Nash

Designed and typeset by Saxon Graphics Ltd, Derby

Original illustrations © Harcourt Education Limited 2004

Printed and bound in Great Britain by Thomson Litho Ltd, East Kilbride

Acknowledgements
The examination questions are reproduced by kind permission of London Qualifications Ltd, trading as Edexcel. The worked exam questions in the text are taken from the following papers:

Unit PSA4
Page 10: PSA4 January 2000; page 13: PSA4 June 2000; page 30: PSA4 June 2001.

Unit PSA5
Page 50: PSA6i June 2000; page 56: PSA6i June 2001; page 66: PSA6i June 2001; page 69: PSA6i June 2001.

The practice exam questions on pages 42–44, 71–74 and 84–85 are taken from the papers specified at the end of each question. Note: London Qualifications Ltd accepts no responsibility whatsoever for the accuracy or method of working in the answers given.

PSA6 worked example 2, page 82 is reproduced by kind permission of DJ Hustings from *Comprehension and Data Analysis Questions in Advanced Physics*, published by John Murray, 1975.

Every effort has been made to contact copyright holders of material reproduced in this book. Any omissions will be rectified in subsequent printings if notice is given to the publishers.

Contents

Introduction – How to revise

This revision guide is written to help you prepare for the A2 exams for the Salters Horners Advanced Physics (SHAP) course. The syllabus is Edexcel Physics (Salters Horners).

In the SHAP course, you study physics within a variety of contexts and applications. The exams will test your knowledge and understanding of the physics concepts and principles, *not* the contexts in which you studied them. The exam questions will ask you to apply this knowledge to particular situations. Some of these might be familiar to you from the course, but many will be unfamiliar.

The SHAP course is designed so that you will meet most key physics concepts several times, gradually building up your experience. In the A2 exams you will sometimes be expected to draw on things you have learned during the AS.

This book summarises the 'bare bones' of the physics content of the course, stripped away from the contexts in which you learned it. This is the content on which examiners can set questions. It is listed in the syllabus specification as 'Learning outcomes' and in your textbook as 'Achievements'.

The **content summary** includes some equations, as these are a convenient way to express many things in physics. **Boxed equations** need to be learned. Others are either given to you at the back of the exam paper or can be derived from those memorised or given. However, equations are *not* the most important aspect of physics.

When revising, try to make sure you understand the topic. If you are unsure of anything, go back to your textbook or ask your teacher for help. Questions and activities at the end of each chapter provide activities to help you, for example making summary charts and mind maps. Ask your teacher and discuss the ideas with other students who are revising the same topic. Talking about ideas and explaining them to yourself and to other people really does help you to make sense of them and fix them in your mind.

At the start of each unit in this revision guide there is a short **introduction** which tells you what is covered by that unit.

Within each unit the content is divided into short spreads of one to four pages, which each revise a section or sections of your textbook. The spreads end with **quick check questions** to help you test your understanding. Try to do them before consulting the **answers** towards the back of the book.

You will find **practice exam questions** from past papers at the end of each unit. Try these only when you are fairly confident that you have revised a unit thoroughly. The outline **answers** at the back of the book indicate points that examiners award marks for. These are *not* model answers!

When answering exam questions, always be careful with details such as units, and using a sharp pencil for graphs. Not doing these could easily lose you marks. Some questions ask you to 'explain' or 'describe' something. This can be more difficult than doing a calculation. Read your answers through afterwards to see if they make sense to someone who does not already know the answer. The **worked examples** and **worked exam questions** in the text are designed to show you the sorts of things the examiners look for.

Unit PSA4: Moving with Physics

This unit tests:

- **Transport on Track (TRA)**
- **The Medium is the Message (MDM)**
- **Probing the Heart of Matter (PRO)**

In this unit, some areas of physics are studied in more than one place – for example, capacitors are met in TRA and again in MDM. To help you bring the ideas together as part of your revision, this part of the revision guide is structured around the physics content rather than following the order of your A2 textbook. The heading of each spread indicates the physics content covered and there is a reference to the relevant section(s) of your textbook where you will find further details.

Momentum conservation

TRA Section 5.2, PRO Section 3.4

Defining momentum

The *momentum*, p, of an object of mass m is

$$p = mv$$

where v is the velocity of the object. Velocity is a *vector* (i.e. it has a direction), therefore momentum is also a vector.

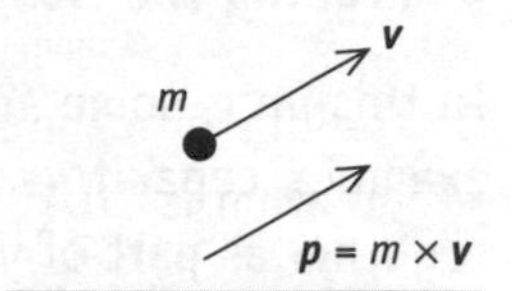

The SI units of momentum are kg m s^{-1}, or N s; 1 kg m s^{-1} = 1 N s.

✓ Quick check 1, 2

Conservation of momentum

The law of *conservation of momentum* is a fundamental law of physics. Provided no external force acts, the total momentum of a system of objects *always* remains constant.

Worked example

A stationary uranium-238 nucleus emits an alpha particle (a helium-4 nucleus) at 1.4×10^7 m s^{-1}. Calculate the recoil speed of the daughter nucleus (thorium-234). (1 u = 1.67×10^{-27} kg)

Step 1 Draw a before-and-after diagram to help you visualise the situation.

before

after

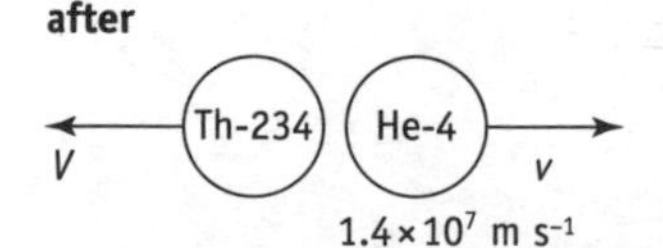

Step 2 Write down what you know:

mass of alpha particle = m = 4 u; mass of thorium-234 = M = 234 u

speed of alpha particle = v = 1.4×10^7 m s^{-1}; recoil speed of thorium-234 = V

Step 3 Use the law of conservation of momentum:

momentum after emission = momentum before emission

As the uranium is at rest, its momentum before emission is zero, so

momentum after emission = $mv + MV = 0$

$$V = \frac{-mv}{M} = \frac{-(4\ \mathrm{u} \times 1.4 \times 10^7\ \mathrm{m\ s^{-1}})}{(234\ \mathrm{u})} = 2.4 \times 10^5\ \mathrm{m\ s^{-1}}$$

It is helpful to choose symbols for the quantities you are using and list their values.

The mass of uranium-238 is 238 u (u = atomic mass unit). Provided you use the same units for mass throughout a momentum calculation, you don't have to convert masses into kilograms.

By convention we take directions to the right as positive, to the left as negative.

Momentum is a vector quantity, so if an event occurs in two or three dimensions then you may have to solve a question by using components or making a scale drawing.

Worked example

A moving lambda particle decays into a proton and a pion, which separate at an angle of 49° as shown in the diagram. The momentum of the proton is 6.0×10^{-18} kg m s^{-1}, and the momentum of the pion is 5.0×10^{-18} kg m s^{-1}. Find by scale drawing the magnitude of the momentum of the lambda particle.

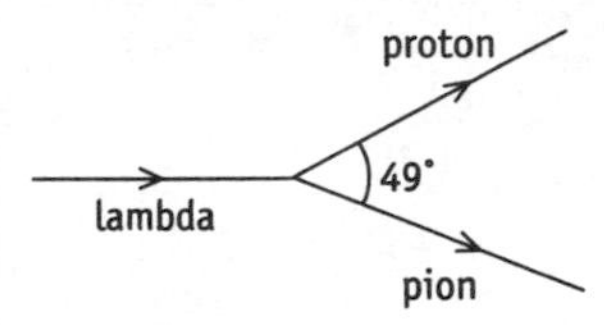

In the drawing the line representing the lambda's momentum is 101 mm long, so its momentum is 10.1×10^{-18} kg m s^{-1}.

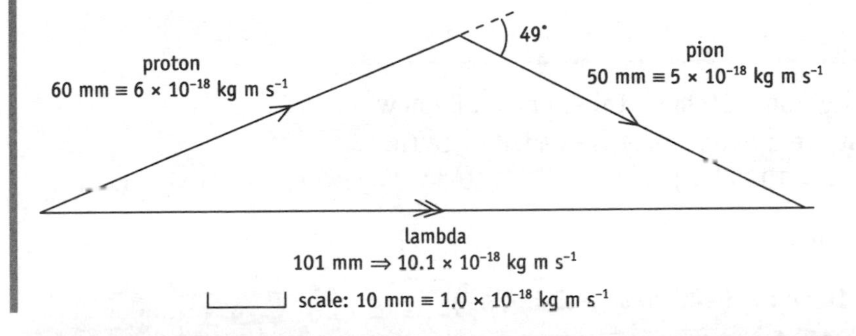

Choose a simple scale which will allow you to make a drawing that fills as much of your available space as possible.

✓ Quick check 3–5

? *Quick check questions*

1. A ball with mass 100 g is travelling at 25 m s^{-1}. What is its momentum in kg m s^{-1}?
2. Show that 1 kg m s^{-1} is equivalent to 1 N s.
3. A truck of mass 4.0 tonnes runs into the back of a stationary 32-tonne locomotive. They move off together at 4.0 m s^{-1}. How fast was the truck travelling before the collision?
4. A rugby player of mass 110 kg running at 9.0 m s^{-1} is tackled by another player of mass 85 kg who collides with him at right angles at 8.0 m s^{-1}.
   a Find by scale drawing their combined momentum (magnitude and direction) after the collision.
   b What is their speed as they move along together after the impact?
5. An object is dropped from rest, accelerates to a speed of a few m s^{-1} then comes to rest again as it hits the floor. Does this break the law of momentum conservation? Explain your answer.

Momentum and force

TRA Sections 3.2 and 5.2

The momentum of a single object changes if an unbalanced *force* acts on it.

force = rate of change of momentum

$$F = \frac{\Delta p}{\Delta t}$$

If mass remains unchanged, then

$$F = m \times \frac{\Delta v}{\Delta t}$$

✓ Quick check 1, 2

Worked examples

1 A car with mass 2.0 t changes its velocity from +30 m s^{-1} to –20 m s^{-1}. By how much does its momentum change? What average net force is needed to cause this change over a time of 25 s? (1 tonne = 1000 kg.)

Step 1 Calculate the change in momentum.

$$\Delta p = p_2 - p_1 = mv_2 - mv_1 = 2.0 \times 10^3 \text{ kg} \times (-20 \text{ m s}^{-1} - 30 \text{ m s}^{-1})$$
$$= -1.0 \times 10^5 \text{ kg m s}^{-1}$$

Take care with signs.

Step 2 Force is the rate of change of momentum, so

$$F = \frac{\Delta p}{\Delta t} = -\frac{1.0 \times 10^5 \text{ kg m s}^{-1}}{25 \text{ s}} = -4.0 \times 10^3 \text{ N}$$

The negative sign shows that the force acts in the opposite direction to the original motion.

2 The graph shows the change of momentum with time of a lorry as it sets off. Find the net forward force accelerating the lorry at time 15 s.

$$\textbf{force = rate of change of momentum} = \frac{\Delta p}{\Delta t}$$

$\Delta p/\Delta t$ is the gradient of the momentum–time graph, so draw a tangent to the graph at t = 15 s.

$$F = \text{gradient of tangent} = \frac{(30-6) \text{ kN s}}{(31-0) \text{ s}} = 0.77 \text{ kN}$$

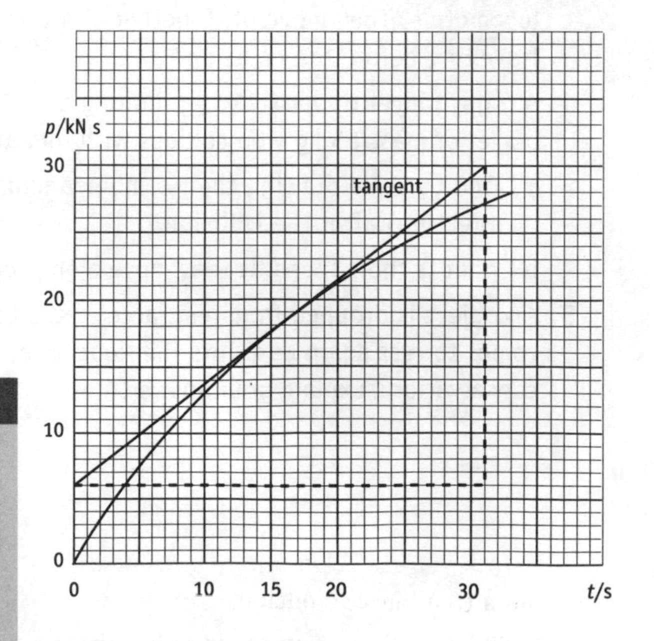

? Quick check questions

1 A railway wagon with mass 20 tonnes is moving at +15 m s^{-1}.

a What is its momentum in kg m s^{-1}?

b If it is braked to a standstill in 10 s, what average net force is acting on it?

2 A model rocket fires exhaust gas backwards at 35 m s^{-1}. If it ejects gas at a rate of 20 grams per second, calculate the forward force on it.

Momentum and kinetic energy

TRA Section 5.3, PRO Section 3.4

The kinetic energy, E_k, of an object is related to its momentum. For speeds much less than *c* (the speed of light)

$$E_k = \frac{1}{2}mv^2 \quad \text{and} \quad p = mv$$

where *v* is the object's speed and *m* its mass.

A little algebra gives

$$E_k = \frac{p^2}{2m}$$

✓ Quick check 1–3

Worked example

A rifle of mass 2 kg fires a bullet of mass 8 g. Compare the kinetic energies of the bullet and the recoiling rifle.

Step 1 The bullet and rifle have equal momentum in opposite directions. *For a given momentum*, the more massive object has smaller kinetic energy:

$$\frac{E_{k\ bullet}}{E_{k\ rifle}} = \frac{(p^2 / 2m_{bullet})}{(p^2 / 2m_{rifle})} = \frac{m_{rifle}}{m_{bullet}}$$

If asked to 'compare ...' two quantities you should find their ratio.

Step 2 Substitute values:

$$\frac{m_{rifle}}{m_{bullet}} = \frac{2000\ g}{8\ g} = 250$$

So the bullet has 250 times as much kinetic energy as the rifle.

? Quick check questions

1 Starting from $p = mv$ and $E_k = \frac{1}{2}mv^2$, derive an expression for *p* in terms of E_k and *m*.

2 For each of the following statements, say whether it applies to **A** momentum, **B** kinetic energy, **C** both, or **D** neither.

- **a** Is a vector.
- **b** Is a scalar.
- **c** Has SI units of J.
- **d** Has SI units of N.
- **e** Is directly proportional to mass.
- **f** Is directly proportional to speed.

3 A stationary uranium-238 nucleus decays into an alpha particle (mass 4 u) and a thorium-234 nucleus.

- **a** Which of these two particles has the greater kinetic energy?
- **b** Compare the kinetic energies of the two particles.

Energy and momentum in collisions

TRA Section 5.3, PRO Section 3.4

When object collide, *momentum* is *always conserved* if there are no external forces acting.

Kinetic energy is *not always conserved*.

- If the total kinetic energy of all the objects is the same after the event as it was before, then kinetic energy is also conserved. This is called an *elastic collision*.
- If there is an overall loss of kinetic energy, then this is an *inelastic collision*.

Perfectly elastic collisions can occur between molecules or smaller particles.

A collision between larger objects is never perfectly elastic if the objects touch one another. There is always some heating of the objects and their surroundings, and a loss of kinetic energy. Sometimes the loss of kinetic energy is very small and you can treat the collision as elastic to make the calculations easier. But sometimes a large fraction of the initial kinetic energy can be lost, or even all of it.

Worked examples

1 An empty railway wagon with mass 8.0 t moving at 4.0 m s^{-1} collides with a stationary loaded wagon with mass 32.0 t. Calculate the speed at which the two wagons move after coupling and the fraction of the initial kinetic energy lost.

Step 1 Consider the momentum before and after the collision.

momentum before = momentum after

$$mu = (M + m)v$$

where m = mass of empty wagon = 8.0 t, u = speed of empty wagon = 4.0 m s^{-1}, M = mass of loaded wagon = 32.0 t, v = speed after coupling.

When analysing a collision it is best to start by considering the momentum unless you already know that it is perfectly elastic.

Step 2 Substitute values and solve for v:

$$v = \frac{mu}{(M+m)} = \frac{8.0\ \text{t} \times 4.0\ \text{m s}^{-1}}{40.0\ \text{t}} = 0.80\ \text{m s}^{-1}$$

The wagons move off at 0.80 m s^{-1}.

Step 3 Now calculate the initial and final kinetic energies:

$$\text{initial } E_k = \tfrac{1}{2}mu^2 = \tfrac{1}{2} \times 8.0 \times 10^3\ \text{kg} \times (4.0\ \text{m s}^{-1})^2 = 6.4 \times 10^4\ \text{J}$$

$$\text{final } E_k = \tfrac{1}{2}(M + m)v^2 = \tfrac{1}{2} \times 4.0 \times 10^4\ \text{kg} \times (0.8\ \text{m s}^{-1})^2 = 1.3 \times 10^4\ \text{J}$$

Step 4 Use these values to find the loss in kinetic energy:

$$\text{loss of } E_k = 6.4 \times 10^4\ \text{J} - 1.3 \times 10^4\ \text{J} = 5.1 \times 10^4\ \text{J}$$

$$\text{fraction lost} = \frac{5.1 \times 10^4 \text{ J}}{6.4 \times 10^4 \text{ J}} = 0.80 \ (= 80\%)$$

✓ Quick check 1

2 The diagram shows a proton colliding with another proton that was initially at rest. After the collision the two particles move apart at right angles. Show that the collision conserved both momentum and kinetic energy.

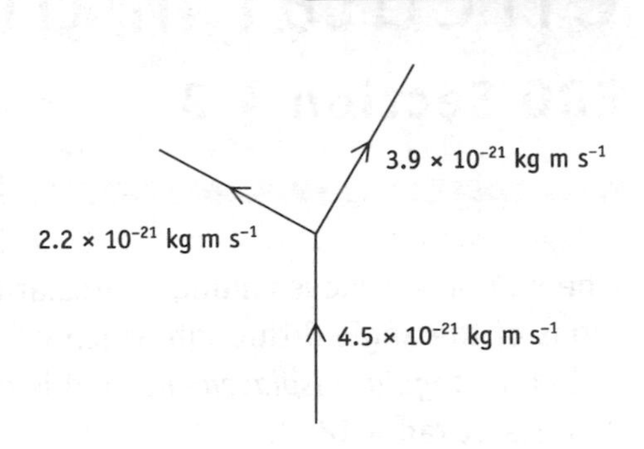

Step 1 Consider the momentum.

One method is to draw a scale diagram like that shown here, where the two sides at right angles have length 22 mm and 39 mm. The measured length of the third side is 45 mm so the total momentum is 4.5×10^{-21} kg m s^{-1} as required.

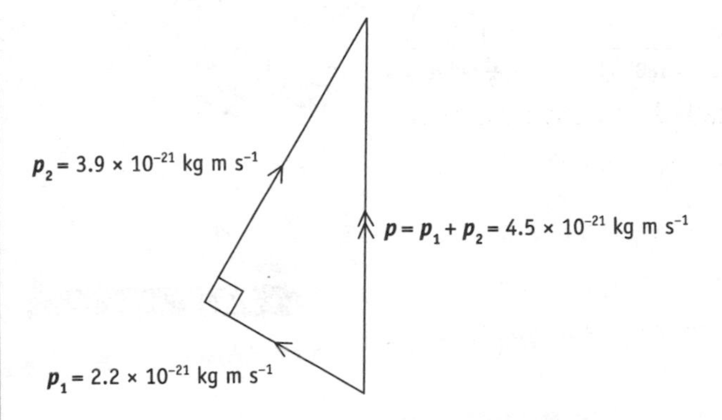

Another method is to apply Pythagoras to find the length of the hypotenuse of the vector triangle. As the two tracks are at right angles, we have

$$p_1^2 + p_2^2 = (2.2 \times 10^{-21} \text{ kg m s}^{-1})^2 + (3.9 \times 10^{-21} \text{ kg m s}^{-1})^2$$
$$= 20 \times (10^{-21} \text{ kg m s}^{-1})^2$$

$$\text{hypotenuse} = \sqrt{20} \times (10^{-21} \text{ kg m s}^{-1}) = 4.5 \times 10^{-21} \text{ kg m s}^{-1}$$

as required.

Step 2 Consider the kinetic energy. We have already shown that

$$p^2 = p_1^2 + p_2^2$$

Since both particles have the same mass, and $E_k = p^2/2m$, it follows that the sum of the two kinetic energies after the collision is equal to the initial kinetic energy of the moving proton.

? Quick check question

1 A car of mass 800 kg travelling at 15.0 m s^{-1} collides head-on with a truck of mass 2000 kg travelling at 5.0 m s^{-1} in the opposite direction. They stick together after the collision. Calculate the fraction of the initial kinetic energy lost in the impact.

Circular motion

PRO Section 4.3

When an object moves along a circular path, it can be useful to state its position in terms of the angle θ through which it has moved from its starting position. This is called its *angular displacement*, and is often measured in radians (rad), rather than degrees. π rad = 180°.

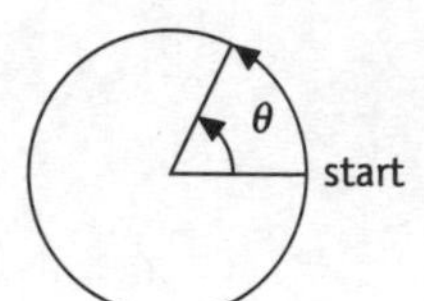

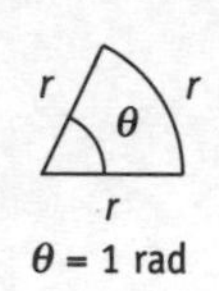

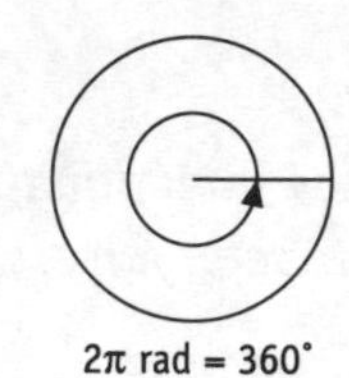

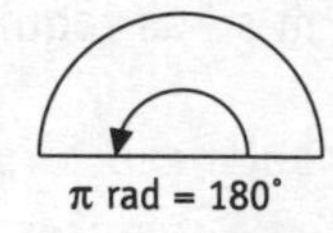

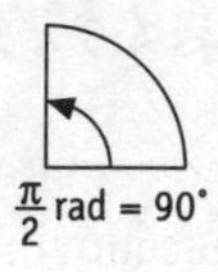

To convert from degrees to radians multiply by $\pi/180$.

To convert from radians to degrees multiply by $180/\pi$.

✓ Quick check 1, 2

Angular velocity and linear speed

Angular velocity, ω, is the angle swept out by a radius per unit time. With θ in radians, ω has units rad s^{-1} or just s^{-1}.

$$\omega = \Delta\theta/\Delta t$$

Angular velocity is related to the number of revolutions per second, i.e. the *frequency*, *f*.

$$\omega = 2\pi f$$

The time *T* for one revolution is given by

$$T = \frac{1}{f} = \frac{2\pi}{\omega}$$

The linear speed *v* of an object moving in a circle is related to its angular velocity:

$$v = r\omega$$

✓ Quick check 3

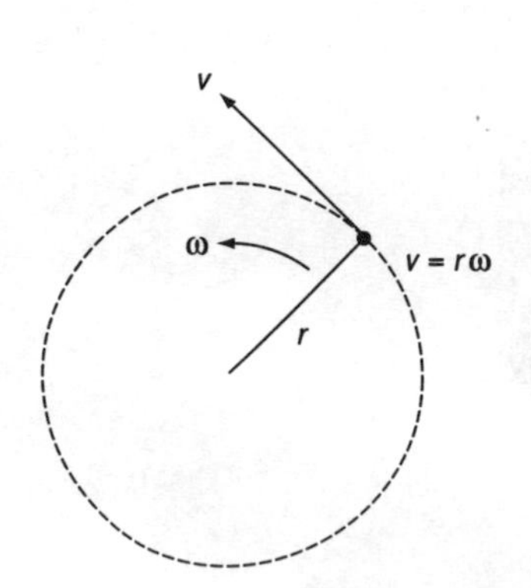

Centripetal force

When an object moves in a curved path, its velocity vector has direction along the tangent to the circle. Even if the speed is constant, the velocity is changing because its direction is changing. Thus it is accelerating.

The small arrow labelled Δv shows how the velocity vector changes from one position to the next. This arrow shows the direction of the change in velocity. It is directed towards centre of the circle; there is a *centripetal acceleration*.

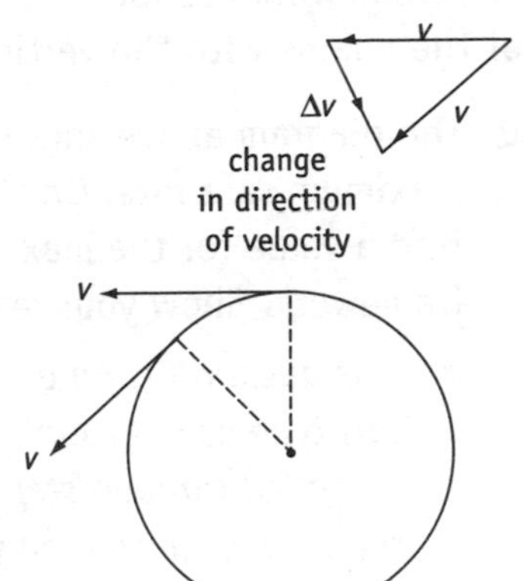

The size of the centripetal acceleration, a, is given by

$$a = \frac{v^2}{r} = r\omega^2$$

An object that is accelerating requires an unbalanced force in the direction of the acceleration. So an object moving in a circle requires a *centripetal force*.

From $F = ma$, the size of the centripetal force, F, is given by

$$F = \frac{mv^2}{r} = mr\omega^2$$

where m is the object's mass.

✓ Quick check 4–6

The diagram shows two examples of how a centripetal force arises.

lift

lift

to centre of circle

A plane banks to follow a curved path.
The horizontal components of the lift forces push the plane towards the centre of the circle.

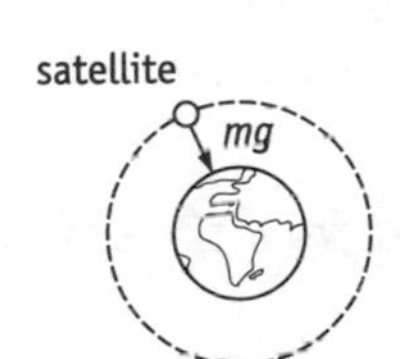

A satellite orbits the Earth.
The Earth's gravitational pull on the satellite is directed towards the centre of the Earth.

Worked example

A light aircraft of mass 500 kg is flying along a curved path of radius 2.0 km at a steady speed of 120 m s^{-1}. Calculate its centripetal acceleration and the centripetal force needed to produce this acceleration.

Step 1 Calculate the acceleration:

$$a = \frac{v^2}{r} = \frac{(120\ \text{m s}^{-1})^2}{2000\ \text{m}} = 7.2\ \text{m s}^{-2}$$

Step 2 Use $F = ma$.

$$F = ma = 500\ \text{kg} \times 7.2\ \text{m s}^{-2} = 3.6 \times 10^3\ \text{N}$$

Worked exam question†

One of the rides at a theme park has a number of chairs, each suspended from a pair of chains from the edge of a framework. The framework rotates so that the chairs swing outwards as they move round in circles.

The framework has radius 4.0 m. The chains are 5.0 m long. For safety, the angle θ of the chains with the vertical must not go above 60°.

Q The diagram above shows a chair swung outwards as the framework rotates at the maximum safe rate. On the diagram draw the forces acting on the chair. Hence find a value for the maximum safe rate of rotation (angular velocity) of the framework. Show your reasoning clearly. [6]

- *This question is an example of an* unstructured problem. *You are likely to find this type of question near the end of the PSA4 or PSA5 test papers. Rather than being led through step by step, you are perhaps given a starting point and are then expected to plan your own way through the problem.*
- *When answering this type of question it is very important that you set out your reasoning clearly and show all your working. You will gain marks if some of this is correct, even if you do not complete the problem. But if the examiner cannot understand what you are doing, you will not get marks.*

A

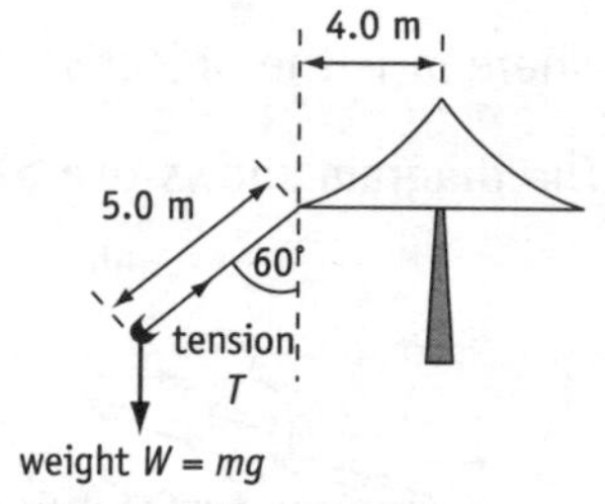

- *Notice that there are just two forces acting: the weight of the chair and the tension in the chain. Centripetal force (or acceleration) should not be included in a force diagram. The centripetal force is provided in this case by a component of the tension; it is not a separate force.*
- *Make sure you draw the force vectors so that they start from the object on which the forces are acting.*
- *This example shows one way to work through the problem. You might think of slightly different ways. Provided the physics is correct, any route through the problem can gain full marks.*

Step 1 Consider the vertical forces.

$$\textbf{vertical component of tension} = T\cos\theta = T\cos 60°$$

Net vertical force on chair = 0 so

$$T\cos\theta = mg \quad \textbf{(1)}$$

> You sometimes need to use earlier work in PSA4 questions. Here you need to use work from PSA1 about components of forces.

Step 2 Consider the horizontal force:

$$\textbf{horizontal component of tension} = T\sin\theta = T\sin 60°$$

> Choose suitable symbols to represent the quantities you are dealing with, e.g. T for the tension.

Step 3 The horizontal component of T produces the centripetal acceleration.

$$T\sin\theta = \frac{mv^2}{r} = mr\omega^2 \quad \textbf{(2)}$$

> As you are asked for angular velocity, use the expression for centripetal acceleration that includes ω.

†The papers the worked exam questions in this guide are taken from are listed on the imprint page at the front of this book.

Step 4 Combine the equations using algebra. Divide equation (2) by equation (1).

$$\frac{T\sin\theta}{T\cos\theta} = \frac{mr\omega^2}{mg}$$

Then cancel:

$$\tan\theta = \frac{r\omega^2}{g}$$

and rearrange to make ω the subject:

$$\omega^2 = \frac{g\tan\theta}{r}$$

Step 5 Calculate a value for radius r:

$$r = 4.0\text{ m} + 5.0\text{ m} \times \sin 60° = 8.33\text{ m}$$

Step 6 Put the values into your expression for ω:

$$\omega^2 = \frac{9.81\text{ m s}^{-2} \times \tan 60°}{8.33\text{ m}} = 2.04\text{ s}^{-2}$$

so ω = **1.43 rad s^{-1}.**

Numbering your equations makes it easy to refer to them and hence to explain what you are doing.

? *Quick check questions*

1. What are the following angles in radians? 360°, 180°, 90°, 60°, 45°.
2. What are the following angles in degrees? 1 rad, 0.25 rad, π rad, π/4 rad.
3. A drill bit with diameter 6.0 mm rotates at 720 rev min^{-1}. Calculate **a** its angular speed and **b** the linear speed at the edge of the bit.
4. The gravitational acceleration near the Moon's surface is 1.6 m s^{-2} and its radius is 1.8×10^6 m. Calculate the speed of a satellite orbiting the Moon close to its surface.
5. In a hammer-throwing event, an athlete whirls round a hammer of mass 4.0 kg at arm's length in a circle of radius 1.2 m. The athlete completes one revolution in 1.4 s. Find the speed of the hammer and hence the tension in the thrower's arms.
6. Using the terms *centripetal force* and *centripetal acceleration*, explain what happens if you whirl an object around on the end of a string, then let go of the string.

Digital signals

MDM Sections 2.2, 2.4 and 2.5

An **analogue** system uses a voltage that varies continuously. An analogue system uses simple electronics, so is cheap and easy to make.

A **digital** system handles information as a sequence of numbers. Each number is represented by a string of *bits* (1s and 0s, or ons and offs). A digital system has several advantages over an analogue system:

- even if individual 1 or 0 signals are distorted, they can still be interpreted perfectly, therefore all noise (uncertainty) can be eliminated;
- binary signals can be processed directly by a computer;
- it is possible to build 'checking' digits into the system to identify errors.

Analogue to digital conversion

The diagram shows the steps involved in **analogue to digital (A to D) conversion**, also known as **pulse code modulation (PCM)**.

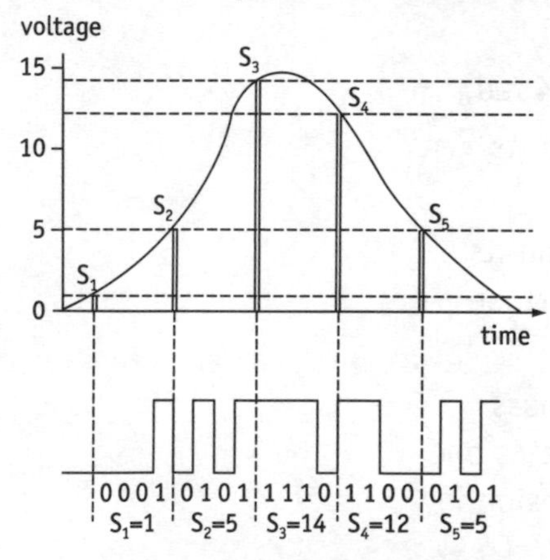

- The analogue voltage is **sampled** (measured) at regular time intervals (S_1, S_2, etc.).
- Each voltage value is assigned to the nearest **quantum level**.
- Each quantum level corresponds to a unique binary number.
- The original information emerges coded as a series of bits.

Make sure you can jot down these steps from memory.

Sampling rate and quantisation

The **sampling rate** (frequency) must be at least twice the highest frequency you want to transmit, otherwise *aliasing* occurs and the original signal cannot be reconstructed. Most humans cannot hear sounds with frequencies above 22 kHz, so CDs sample at 44 kHz.

More bits per sample gives more accurate **quantisation**, but takes more time to transmit and space to store. For example, using 4 bits to encode a sample gives 2^4 (= 16) quantum levels, as in the diagram. Real systems usually have smaller spaces between their quantum levels for low voltages. This improves the quality of reproduction, and is called **companding**.

✓ Quick check 1

Worked exam question

It has been said that using high-speed digital signal transmission, the complete texts of all Shakespeare's plays can be transmitted in about one quarter of a second. A typical *Complete Plays of Shakespeare* in book form has about 1100 pages.

Q Make an order-of-magnitude estimate of the bit rate of the signal that would be needed. State any quantities that you are estimating and show your reasoning clearly. [5]

A **Step 1** Suppose each page has 100 lines, each line has about 20 words and each word has about five letters on average. Then

no. of characters in the book $= 1100 \times 100 \times 20 \times 5 = 1.1 \times 10^7$

- *Say briefly what each number represents.*
- *You would get full marks for any reasonable values (e.g. saying there are 80 lines on a page would be fine, but saying each page has just 10 lines would not be sensible).*
- *Choose round numbers to make the arithmetic simple.*

Step 2 Estimate how many bits are needed for each character.

There must be at least 52 characters (26 letters, and each can be capital or lower-case). With other characters representing punctuation marks and blank spaces, a reasonable estimate might be about 65 different characters altogether.

Each character must be represented by a different binary number in the range 0–64. As $64 = 2^6$, we need 6 bits to encode all the different characters.

- *If you estimated 75 characters you would need 7 bits, which would actually be enough for 128 different characters ($2^7 = 128$).*

Step 3 Use your estimates to find the number of bits needed.

total no. of bits = no. of characters × no. of bits to encode each character

$$= 6 \times 1.1 \times 10^7 = 6.6 \times 10^7$$

Step 4 Calculate how many bits per second need to be sent.

Sending 6.6×10^6 bits in 0.25 s is equivalent to

$$4 \times 6.6 \times 10^7 \text{ bits per second} = 2.64 \times 10^8 \text{ bits per second}$$

$$= 3 \times 10^8 \text{ bits per second}$$

- *When asked for an order of magnitude estimate you should round your final answer to just one significant figure.*

? Quick check question

1 A particular system encodes each sample using 8 bits. How many quantum levels are there?

Sending signals

MDM Sections 2.1 and 3.1

Analogue signals can be carried on radio or light waves that have undergone **amplitude modulation** (AM) or **frequency modulation** (FM).

AM uses a carrier wave of a fixed frequency. The signal is combined with the carrier so that the carrier *amplitude* varies exactly like the signal voltage shown in the diagram.

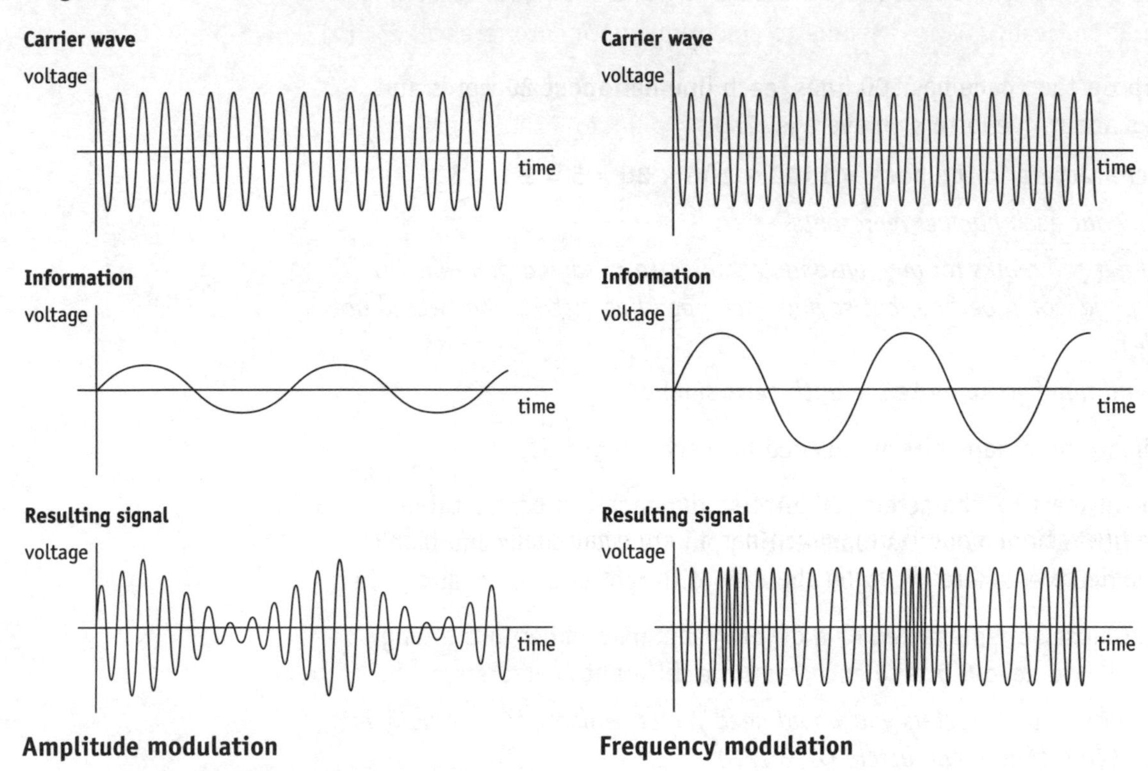

Amplitude modulation **Frequency modulation**

In FM, the signal is combined with a carrier wave in such a way that the *frequency* of the carrier varies exactly like the signal voltage, as shown here.

Transmitting multiple signals

Carrier waves with many different frequencies can be sent simultaneously along a single cable or optical fibre. This is called **frequency division multiplexing (FDM)**. Each carrier frequency carries a different signal. They are separated by suitable electronics at the receiving end. Thus for example a single optical fibre brings many TV channels into your house.

The technique of **time division multiplexing (TDM)** is used to pass many *digital* signals along a single fibre or cable at the same time:

- each signal is broken into blocks of equal length (e.g. 8 bits);
- a block from each signal in turn is sent along the fibre;
- at the receiving end the blocks are sorted electronically to reconstitute the original signals.

Both analogue and digital information can be carried:

- by any electromagnetic waves (usually radio) travelling freely through space
- by radio waves along coaxial cable
- by light along an *optical fibre*.

✓ Quick check 1, 2

Optical fibres

Optical fibres have the advantage that:

- light has a higher frequency than radio so can transmit more information per second;
- optical fibres have low mass and use cheap material (glass);
- signals along an optical fibre are immune to electromagnetic interference.

One problem with optical fibre signals is **multipath dispersion**: some rays travel along longer paths than others so a single pulse which is initially sharp becomes smeared out.

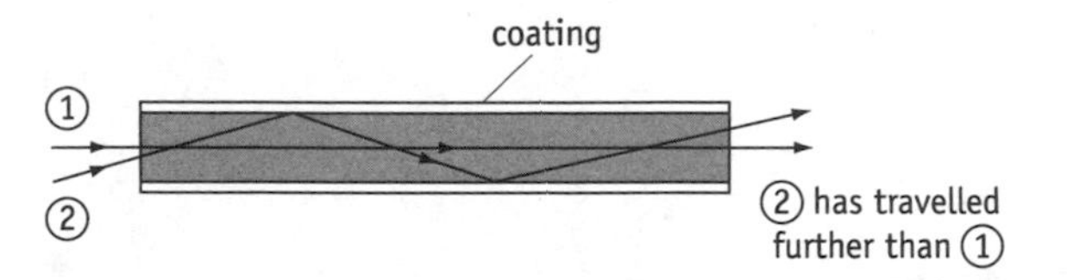

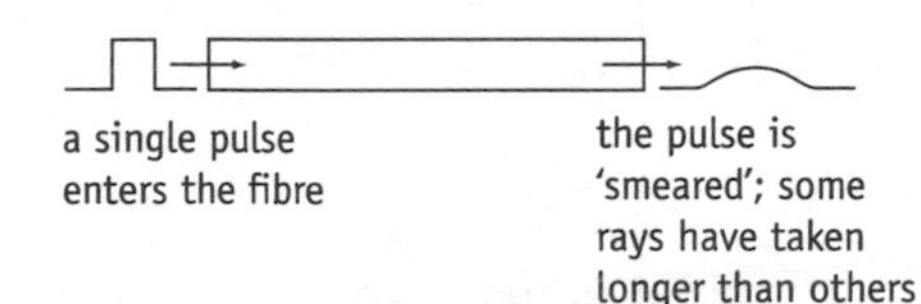

Multipath dispersion can be reduced by using

- a *graded index fibre*, in which the refractive index gradually decreases away from the axis of the fibre, so the different paths travelled will take similar times
- a *single mode fibre* (monomode fibre), which has a very small diameter so that the only available path is directly along the axis of the fibre.

Quick check questions

1. A certain system samples at 44 kHz, and encodes each sample using 8 bits. How many such signals could be sent using TDM along a one gigahertz fibre, i.e. one capable of handling 10^9 bits per second?
2. For each of the following, say whether it can be used with **A** analogue signals, **B** digital signals, or **C** both.
   - **a** amplitude modulation
   - **b** frequency modulation
   - **c** frequency division multiplexing
   - **d** time division multiplexing
   - **e** transmission by radio waves
   - **f** transmission by light along optical fibre.

Capacitors

TRA Section 4.2, MDM Sections 4.1 and 4.2

Capacitors are components used in circuits to store energy and electric charge. They usually consist of two parallel metal plates or foils separated by a thin layer of insulating material.

Behaviour of capacitors

The circuit shown here can be used to demonstrate the behaviour of capacitors.

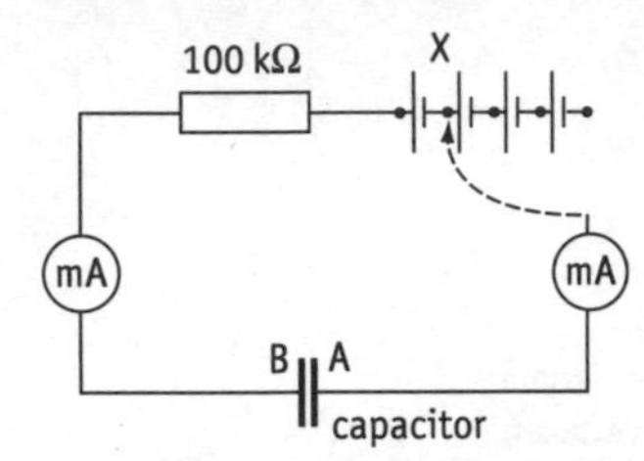

- Start with an uncharged capacitor.
- Touch the flying lead to point X. A burst of current passes through both meters.
- Estimate the amount of charge which has moved by finding the area under the graph of current against time.
- Both meters show the same burst. The same quantity of charge moves off plate A, through the cell, and onto plate B.
- The amount of charge stored, Q, is directly proportional to the potential difference, V, across the capacitor:

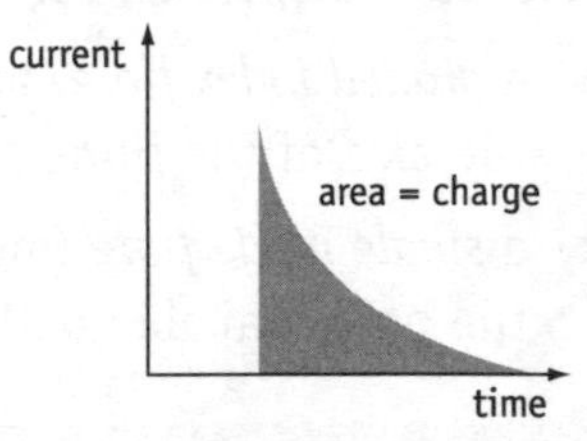

$$Q = CV$$

where C is the capacitance.

Capacitance has SI units of C V^{-1} or farads, F: 1 F = 1 C V^{-1}.

Most practical capacitors have capacitance much less than 1 F:

- 1 μF = 1 microfarad = 10^{-6} F
- 1 pF = 1 picofarad = 10^{-12} F.

Take care not to confuse C for coulomb, the unit of charge, with C the symbol for capacitance.

✓ Quick check 1

Charging a capacitor

The diagram shows three graphs for a charging capacitor. The shapes are all related to one another.

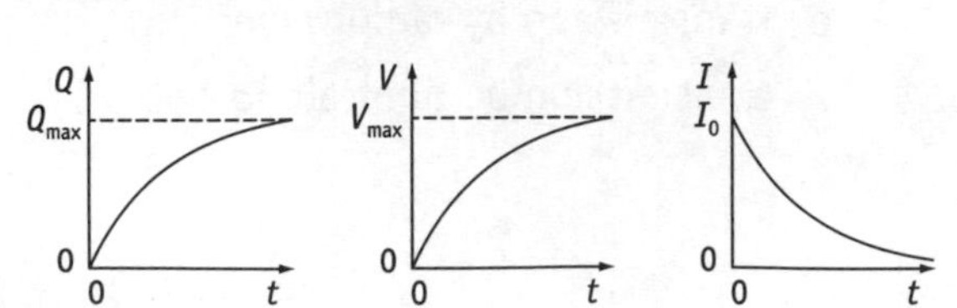

As the stored charge increases, the pd across the capacitor increases.

When the capacitor pd *rises*, the pd across the resistor *falls* because the total pd is always the same as the terminal pd of the power supply.

Since the pd across the resistor falls, the current in the circuit also falls.

Charge builds up on the capacitor rapidly at first, then more slowly.

✓ Quick check 2

Storing energy

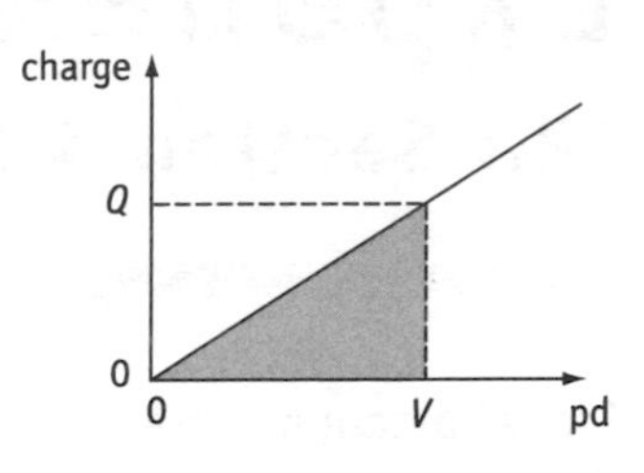

When a capacitor is being charged, *work* is done by the potential difference that pushes charge onto the plates. This means that a charged capacitor stores *energy*. This energy is released when the capacitor is discharged.

energy stored = work done = charge × pd

The energy W stored by a charged capacitor is the triangular area under the graph of charge against pd. Therefore

$$W = \tfrac{1}{2} QV$$

✓ Quick check 3, 4

Worked examples

1 A 600 pF capacitor is connected to a potential difference of 12 V. How much charge and energy does it store?

Step 1 Calculate the charge:

$$C = 600 \text{ pF} = 600 \times 10^{-12} \text{ F}$$

$$Q = CV = 600 \times 10^{-12} \text{ F} \times 12 \text{ V} = 7.2 \times 10^{-9} \text{ C}$$

Step 2 Use this value of Q to find the energy stored:

$$W = \tfrac{1}{2} QV = \tfrac{1}{2} \times 7.2 \times 10^{-9} \text{ C} \times 12 \text{ V} = 4.3 \times 10^{-8} \text{ J}$$

2 A 2.2 μF capacitor stores 3.0 μC of charge. If it discharges through a 1.5 kΩ resistor, what is the initial discharge current?

Step 1 Find the pd across the capacitor:

$$V = \frac{Q}{C} = \frac{3.0 \times 10^{-6} \text{ C}}{2.2 \times 10^{-6} \text{ F}} = 1.36 \text{ V}$$

Step 2 Substitute for V:

$$I = \frac{V}{R} = \frac{1.36 \text{ V}}{1.5 \times 10^{3} \text{ }\Omega} = 9.1 \times 10^{-4} \text{ A } (= 0.91 \text{ mA})$$

? Quick check questions

1 A capacitor stores 40 μC of charge when connected to a 5.0 V supply. **a** What is its capacitance? **b** How much charge will it store when connected to a 20 V supply?

2 This circuit is used for discharging a capacitor. The discharge is shown on the three graphs. Write a few sentences to explain the relationships between the shapes of the graphs.

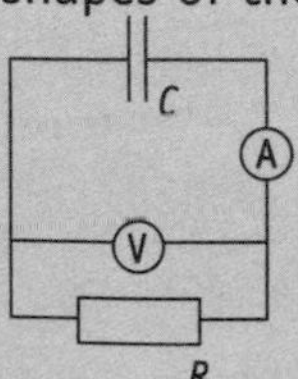

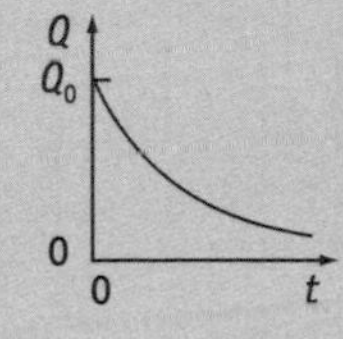

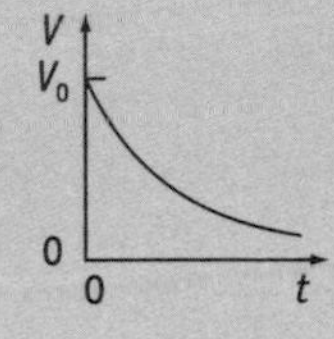

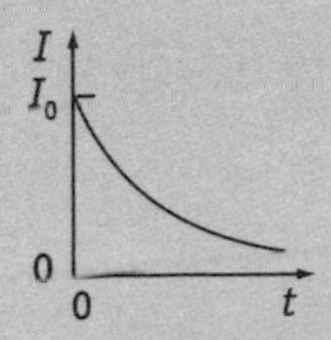

3 Derive expressions for the energy, W, stored by a capacitor in terms of **a** C and V; **b** C and Q.

4 A capacitor stores 10 mJ of energy when connected to a 100 V supply.
a What is its capacitance?
b How much energy will it store when connected to a 200 V supply?

Exponential decay

TRA Section 4.2, MDM Section 3.2

Signal attenuation and capacitor discharge are both examples of **exponential decay.**

Time constant

When a charged *capacitor* is *discharged* through a *resistor*, the charge Q left on the capacitor, the pd V across it and the current I all fall with similar exponential curves.

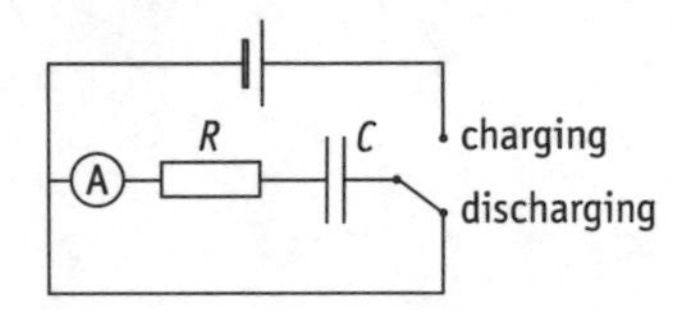

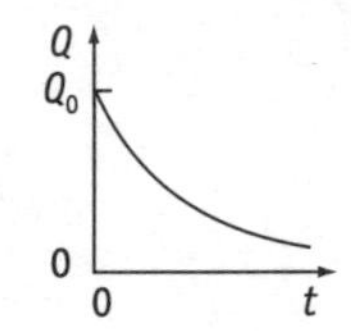

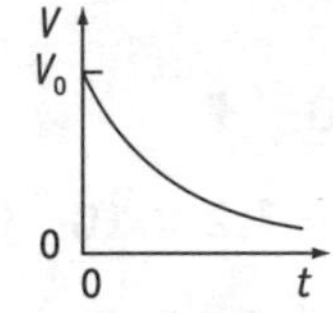

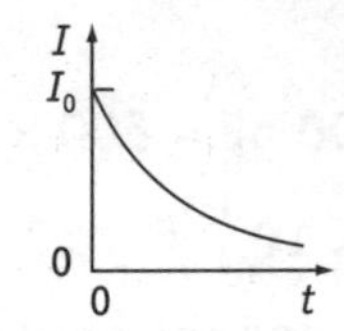

The **time constant** τ of a circuit is the time for the charge, pd or current to fall to 1/e of its initial value, which is about 37% or 0.37.

$$\tau = RC$$

With R in ohms and C in farads, τ is a time in seconds.

The following equations can be used to find the charge, pd and current at any time during the discharge.

$$Q = Q_0 e^{-t/RC} \quad V = V_0 e^{-t/RC} \quad I = I_0 e^{-t/RC}$$

Q_0, V_0 and I_0 are the charge, pd and current when $t = 0$.

Attenuation coefficient

A signal travelling through any material (e.g. in an optical fibre) is *attenuated*, meaning it gets weaker. The intensity I generally falls exponentially with distance.

$$I = I_0 e^{-\mu x}$$

where I is the intensity at distance x from the start, I_0 is the intensity at $x = 0$, and μ is the **attenuation coefficient**. A large value of μ means a lot of attenuation over a short distance.

The SI units of μ are m^{-1}. The quantity μx has no units.

Take care not to confuse the SI prefix μ (= 10^{-6}) with μ the coefficient.

✓ Quick check 1–5

Worked examples

1 A 20 μF capacitor is charged to 15 V then discharged through a 40 kΩ resistor. Calculate the initial charge on the capacitor and the charge remaining after 4.0 s.

Step 1 Calculate the initial charge:

$$Q_0 = CV_0 = 20 \times 10^{-6}\ \text{F} \times 15\ \text{V} = 3.0 \times 10^{-4}\ \text{C}$$

Step 2 Find t/RC:

$$\frac{t}{RC} = \frac{4.0\ \text{s}}{40 \times 10^3\ \Omega \times 20 \times 10^{-6}\ \text{F}} = 5.0$$

> The quantity t/RC has no units as it is a ratio of two times.

Step 3 Find $e^{-t/RC}$ using the e^x button on a calculator (or inv then ln) and multiply this by Q_0:

$$Q = Q_0 e^{-t/RC} = 3.0 \times 10^{-4}\ \text{C} \times e^{-5.0} = 2.0 \times 10^{-6}\ \text{C}\ (= 2.0\ \mu\text{C})$$

2 A signal along an optical fibre falls to 20% of its initial intensity after 2.0 km. Calculate the attenuation coefficient μ of the fibre.

Step 1 Make μx the subject of the equation.

$$\frac{I}{I_0} = e^{-\mu x}$$

$$\ln\left(\frac{I}{I_0}\right) = -\mu x$$

> To 'undo' an exponential function you need to use the natural logarithm ln.

Step 2 Substitute values.

When $x = 2.0$ km $= 2.0 \times 10^3$ m, we have $I/I_0 = 20\% = 0.2$.

$$\ln(0.2) = -1.6 \quad \text{so} \quad \mu = \frac{-1.6}{-2.0 \times 10^3\ \text{m}} = 8.0 \times 10^{-4}\ \text{m}^{-1}$$

> Take care with signs. $\ln(I/I_0)$ has a negative sign so dividing by $-x$ gives a positive value for μ.

? Quick check questions

1. Calculate the time constant of a circuit which has a 20 μF capacitor and a 5 MΩ resistor.
2. Which has the greater time constant, a circuit which has a 40 μF capacitor and a 2.5 MΩ resistor or one with a 1000 μF capacitor and a 100 kΩ resistor?
3. A capacitor is charged, then discharged through a resistor. It is then charged again to the same pd, and discharged through a resistor of twice the resistance. Sketch two graphs on the same axes showing the charge on the capacitor during the discharge. Explain any similarities and differences between the two curves.
4. A 20 pF capacitor is charged to 100 V, then discharged through a 500 MΩ resistor. Calculate **a** the initial charge on the capacitor, **b** the time constant of the circuit, and **c** the charge left on the capacitor 0.015 s after discharging begins.
5. Ultraviolet radiation crossing our galaxy is attenuated exponentially. The intensity is reduced to 10% of its initial value after a distance of 3×10^{17} m. Calculate the value of μ.

Exponential relationships and log graphs

TRA Section 4.2, MDM Section 3.2

An *exponential relationship* in which y gets smaller as x increases is described by the general equation

$$y = Ae^{-kx}$$

Examples of exponential decay include:

Attenuation of signals $I = I_0\, e^{-\mu x}$	$x \rightarrow x$ (distance along cable, fibre, etc.) $k \rightarrow \mu$ (attenuation coefficient) $y \rightarrow I$ (intensity of signal at distance x) $A \rightarrow I_0$ (intensity at $x = 0$)
Capacitor–resistor circuits $Q = Q_0 e^{-t/RC}$ $V = V_0 e^{-t/RC}$ $I = I_0 e^{-t/RC}$	$x \rightarrow t$ (time) $k \rightarrow 1/RC$ (1/time constant) $y \rightarrow Q$ (charge on capacitor) V (pd across capacitor) I (current in resistor) $A \rightarrow Q_0$ (initial charge) V_0 (initial pd) I_0 (initial current)
Radioactive decay of nuclei $N = N_0 e^{-\lambda t}$ $A = A_0 e^{-\lambda t}$	$x \rightarrow t$ (time) $k \rightarrow \lambda$ (decay constant) $y \rightarrow N$ (number of nuclei remaining) A (activity of sample) $A \rightarrow N_0$ (initial number of nuclei) A_0 (initial activity)

▶▶ *Radioactive decay is covered in the PSA5 unit Reach for the Stars.*

Testing for an exponential relationship

There are two ways to test whether data obey an exponential relationship.

- Plot a graph of y against x.

If the relationship is exponential, then y will change by *equal fractions* for *equal steps* in x.

To find the decay constant from the graph, find the value of x for which y = A/e (= 0.37A). Then k = 1/x.

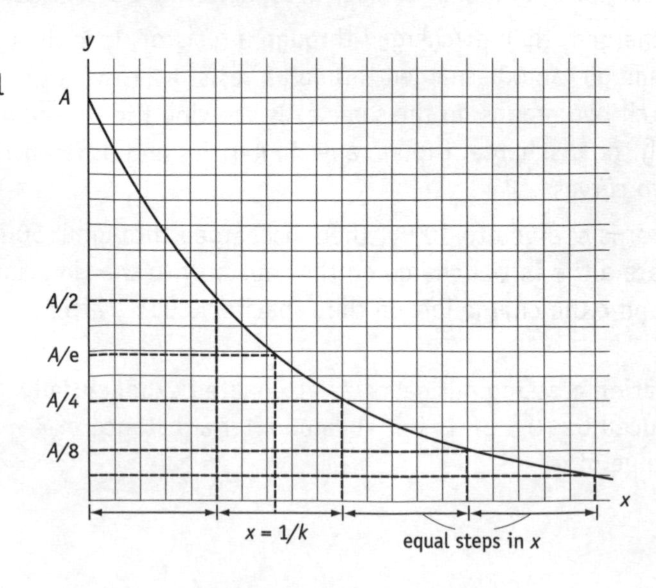

- Plot a graph of ln (y) against x (i.e. a *log–linear graph*)

If the relationship is exponential, then the graph is a straight line, and the gradient of the graph is equal to the decay constant k.

To see why this works, we need some maths. Starting from

$$y = Ae^{-kx}$$

take natural logs of both sides:

$$\ln(y) = \ln(A) - kx \quad \text{or} \quad \ln(y) = -kx + \ln(A)$$

With ln (y) on the vertical axis and x on the x-axis, this has the form of equation for a straight line $y = mx + c$.

> ln (y) is the logarithm to base e of y, also known as the natural logarithm of y. There is a button on your calculator that calculates its value.

> If you are not very mathematical, then just learn to do these operations without worrying what they mean!

Worked example

A student records the voltages across a capacitor discharging through a 470 kΩ resistor. Her results are in columns A and B of the spreadsheet. Show that these results fit an exponential decay for the voltage and find the value of the capacitance C.

	A	B	C
1	time (s)	voltage (V)	*ln (voltage)*
2	0	10	*2.30*
3	10	8	*2.08*
4	20	6.4	*1.86*
5	30	5.1	*1.63*
6	40	4.1	*1.41*
7	50	3.3	*1.19*

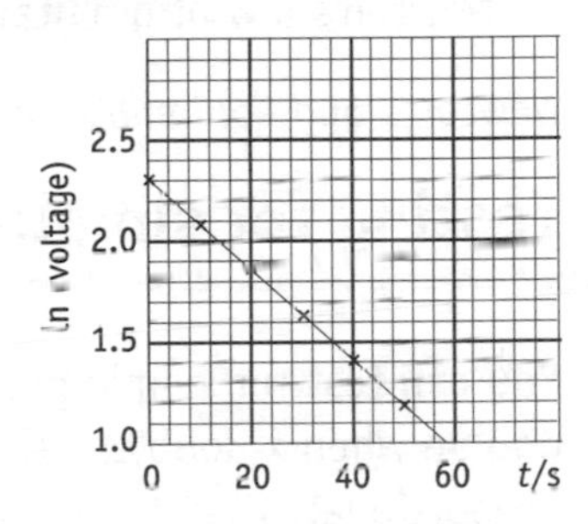

Step 1 Plot ln (voltage) against time. The graph is a straight line, so the relationship is exponential.

Step 2 The gradient of the graph is $-1/RC$:

$$\textbf{gradient} = \frac{(1.0 - 2.3)}{(59\text{ s} - 0\text{ s})} = -0.0220\text{ s}^{-1}$$

$$RC = \frac{1}{0.0220\text{ s}^{-1}} = 45.4\text{ s}$$

$$C = \frac{45.4\text{ s}}{470 \times 10^{3}\ \Omega} = 9.7 \times 10^{-5}\text{ F}$$

Quick check question

✓ Quick check 1

1 The table lists measurements of signal voltage at various positions along a cable. By plotting graphs of voltage against distance and ln (voltage) against distance, show that the data obey an exponential relationship and find the attenuation coefficient.

distance x/km	signal voltage V/V	ln (V)
0	6.00	
100	4.67	
200	3.63	
300	2.83	
400	2.21	
500	1.72	
600	1.34	

Power laws and log graphs

PRO Section 2.5

Two variables x and y are related by a *power law* if they obey an equation of the form

$$y \propto x^p \quad \text{or} \quad y = Ax^p$$

where A and p are constants. p is called the *exponent*.

Do not confuse the word 'exponent' here with anything to do with exponential changes.

Examples of power law relationships include:

- the energy stored in a charged capacitor ($W \propto V^2$)
- the centripetal force required to keep an object in a circular path ($F \propto v^2$ or $F \propto \omega^2$)
- Coulomb's law of electrostatic force between point charges ($F \propto r^{-2}$)
- Newton's law of gravitational force between point masses ($F \propto r^{-2}$)

▶▶ *Newton's and Coulomb's laws are covered in the PSA5 unit Reach for the Stars.*

Testing for simple proportion

You can test for simple proportion ($y \propto x$) quickly by looking at the data. Does y double when x doubles? If the answer is yes, then they are probably directly proportional. Then try a graph of y against x. If the graph is a straight line through the origin, that confirms that they are directly proportional.

Testing for a power law relationship

If you want to confirm that data are described by a power law with a known exponent p, then plot a graph of y against x^p. The graph will be a straight line through the origin and its gradient is equal to the constant A.

You might have a table of values of some variables, and you want to test whether they obey a power law but it is not obvious what the exponent might be.

One method is to plot a graph of log (y) against log (x). This line of maths explains why:

$$y = Ax^p$$

$$\rightarrow \log(y) = \log(A) + p\log(x)$$

If the graph gives a straight line, then

- the relationship is indeed a power law as you suspected,
- the gradient of the line gives the value of p.

For this method, it does not matter what units x and y are in, so long as within a column of data the units are all the same. Nor does it matter what base of logarithms you use, so long as you are consistent. Your calculator can do ln (base e) or log (base 10).

> If you are not very mathematical, then just learn to do these operations without worrying what they mean!

Worked example

Columns A and B show data from a test of Coulomb's law, in which the force between two charged object is measured when they are separated by various distances r. Show that the data obey a power law, and find a value for p.

	A	B	C	D
1	r (mm)	force (unknown units)	*log (r)*	*log (force)*
2	106	8	*2.03*	*0.90*
3	81	16	*1.91*	*1.20*
4	62	32	*1.79*	*1.51*
5	54	53	*1.73*	*1.72*
6	45	72	*1.65*	*1.86*
7	33	110	*1.52*	*2.04*

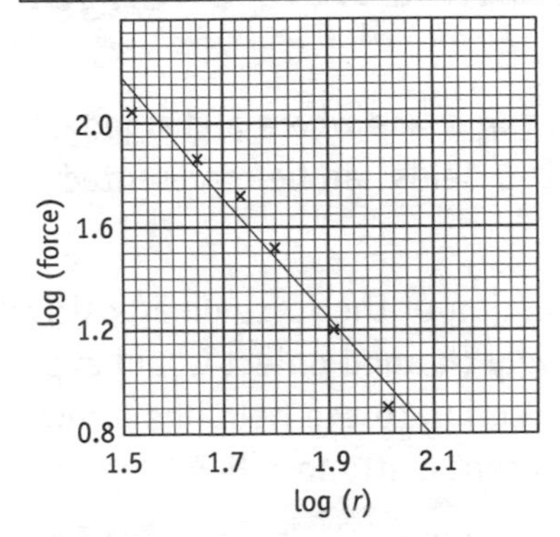

Step 1 The graph of log (force) against log (r) is (reasonably) straight, so the data fit a power law to reasonable accuracy.

Step 2 The gradient of the graph has the value p.

$$\textbf{gradient} = \frac{(0.80 - 2.17)}{(2.09 - 1.50)} = -2.3 \quad \text{so} \quad p = -2.3$$

The expected value here is $p = -2$. However, the scatter in the graph suggests a large experimental uncertainty in p so the data are probably consistent with $p = -2$.

> The working given here is using base 10 logs. A graph plotted with natural logs (base e) would have different values on the axes but the same gradient.

✓ *Quick check 1*

? *Quick check question*

1 The table lists the orbital periods of the Solar System planets and their distances from the Sun. By plotting a log–log graph, show that the data obey a power law of the form $T = Ar^p$ and find the value of p.

planet	**distance from Sun** $r/10^6$ km	**orbital period** T/years
Mercury	57.9	0.241
Venus	108.2	0.612
Earth	149.6	1.00
Mars	227.9	1.52
Jupiter	778.3	11.9
Saturn	1427	29.5
Uranus	2870	84.0
Neptune	4497	165
Pluto	5900	247

Electric forces and fields

MDM Sections 5.2 and 5.3, PRO Section 3.3

Electric forces are produced by objects carrying electric *charge*. There are two types of charge, positive (+) and negative (–). Two like charges (+ + or – –) repel each other; two unlike charges (+ – or – +) attract each other.

Field lines and electric field strength

An *electric field* is a region of space where a charged object experiences a force. Electric fields can be represented diagramatically.

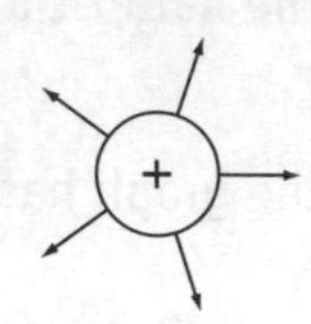

isolated positive charge

A *field line* shows the direction of the force on a positively charged object. A negatively charged object would experience a force in the opposite direction to the field line arrow. Lines closer together represent a stronger field.

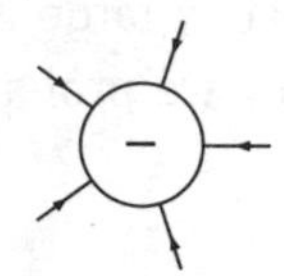

isolated negative charge

To define *electric field strength* E at a point in a field, imagine placing a small positive charge Q there. Measure the force F acting on the charge. Then

$$E = \frac{F}{Q}$$

E has SI units N C^{-1}.

If there is potential difference across two parallel plates, there is a *uniform* electric field between them. If the plates are distance d apart and the potential difference is V, then

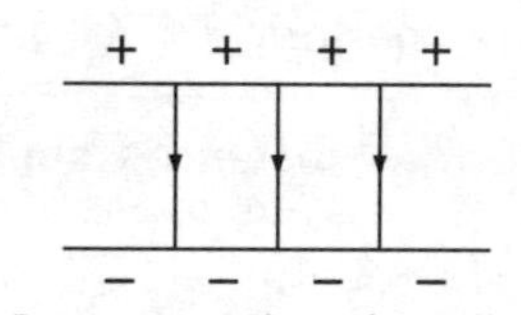

Between two charged, parallel plates, the field is uniform

$$E = \frac{V}{d}$$

This gives V m^{-1} as the SI units of E. N C^{-1} and V m^{-1} are equivalent. You can use either – they mean the same thing.

Quick check 2, 3

Coulomb's law

If two *point* charges, or the centres of two spheres of charge, Q_1 and Q_2 are separated by a distance r, the force F between them is described by *Coulomb's law*:

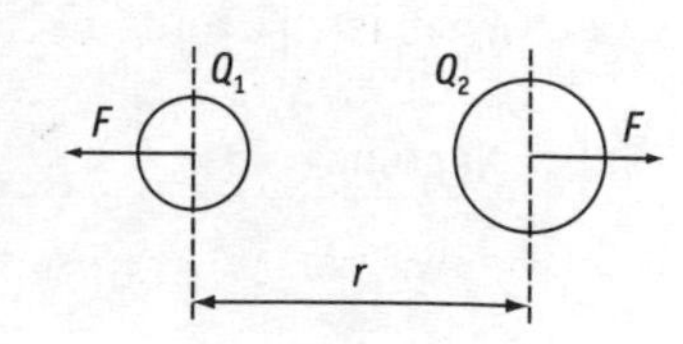

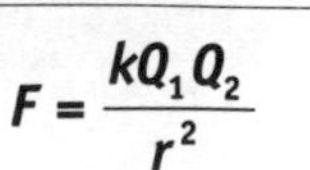

$$F = \frac{kQ_1Q_2}{r^2}$$

where

$$k = \frac{1}{4\pi\varepsilon_0}$$

and ε_0 is the *permittivity of free space*.

- $k = 8.99 \times 10^9$ N m^2 C^{-2}
- $\varepsilon_0 = 8.85 \times 10^{-12}$ F m

Coulomb's law is one example of an *inverse square law*, so called because the force is *inversely* proportional to the *square* of the separation.

✓ Quick check 4

The *electric field strength E* at a distance *r* from a point charge *Q* is given by

$$E = \frac{kQ}{r^2} = \frac{Q}{4\pi\varepsilon_0 r^2}$$

This comes from dividing the Coulomb's law equation by one of the charges.

✓ Quick check 5

Worked example

Calculate the electric field strength at a distance of 0.09 nm from an oxygen nucleus, and the force between an electron and an oxygen nucleus separated by this distance. The oxygen nucleus has charge +8*e* where $e = 1.60 \times 10^{-19}$ C.

Step 1 Calculate *E*:

$$E = \frac{kQ}{r^2}$$

$$= \frac{8.99 \times 10^9 \text{ N m}^2 \text{ C}^{-2} \times 8 \times 1.60 \times 10^{-19} \text{ C}}{(0.09 \times 10^{-9} \text{ m})^2}$$

$$= 1.42 \times 10^{12} \text{ N C}^{-1}$$

Step 2 Use this value to calculate the force *F*:

$$F = eE = 1.42 \times 10^{12} \text{ N C}^{-1} \times 1.60 \times 10^{-19} \text{ C}$$

$$= 2.3 \times 10^{-7} \text{ N}$$

? Quick check questions

1. A proton is in an electric field of strength 1.5 kN C^{-1}. What is the electric force acting on it? (Proton charge $e = 1.60 \times 10^{-19}$ C.)
2. Two parallel metal plates, separated by 15 cm, are connected to the terminals of a 120 V supply. What is the electric field strength between them?
3. A dust particle carrying a charge of 3 μC is between two parallel plates connected to a potential difference of 45 V. If the plates are 25 cm apart, what is the force on the dust particle?
4. What is the force between two protons in an atomic nucleus, separated by 2.0×10^{-15} m?
5. A metal sphere of radius 20 mm carries a charge of 70 μC. What is the field strength at distance 10 mm from the surface of the sphere?

Particles in electric fields

MDM Section 5.3, PRO Section 4.2

Electric fields can be used to accelerate charged particles to high energies.

The electron gun

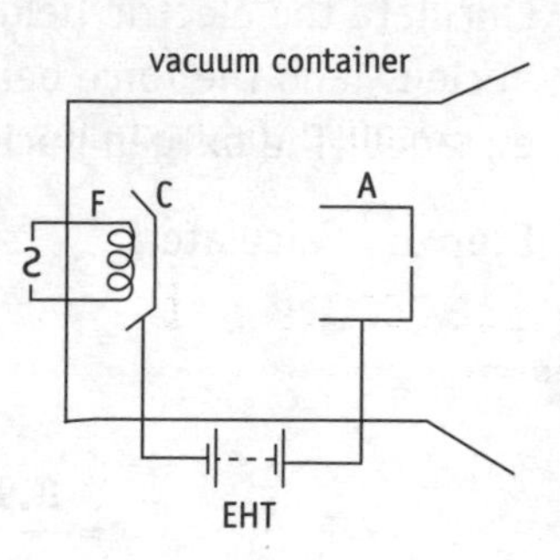

- A metal filament F is heated by passing current through it.
- The filament heats up the cathode C placed close to F.
- Electrons in C gain enough energy to escape from the surface of C (thermionic emission).
- Between the cathode C and the system of anodes A there is a high potential difference V (sometimes called the gun voltage or 'high tension' or EHT).
- The electric field caused by the gun voltage between C and A accelerates the electrons.
- The anodes A are arranged to focus the electrons into a beam (sometimes just a hole in a single anode is enough).
- The components F, C and A are in a vacuum container to prevent gas molecules impeding the beam.

Practise jotting down this summary from memory.

Electrons passing through the gun gain kinetic energy E_k

$$E_k = eV$$

where e is the electron charge ($e = 1.60 \times 10^{-19}$ C).

Provided the speed, v, is much less than the speed of light ($c = 3.00 \times 10^8$ m s^{-1}) then the *electron gun equation* is

$$eV = \frac{1}{2}mv^2$$

Worked example

Electrons are accelerated through a pd of 200 V. What is their speed as they leave the gun? (Mass of electron $m = 9.11 \times 10^{-31}$ kg.)

Step 1 Make v the subject of the gun equation:

$$v^2 = \frac{2eV}{m}$$

Step 2 Substitute values:

$$v = \sqrt{\left(\frac{2 \times 1.60 \times 10^{-19}\ \text{C} \times 200\ \text{V}}{9.11 \times 10^{-31}\ \text{kg}}\right)}$$

$$= 8.4 \times 10^6\ \text{m s}^{-1}$$

Linear accelerator

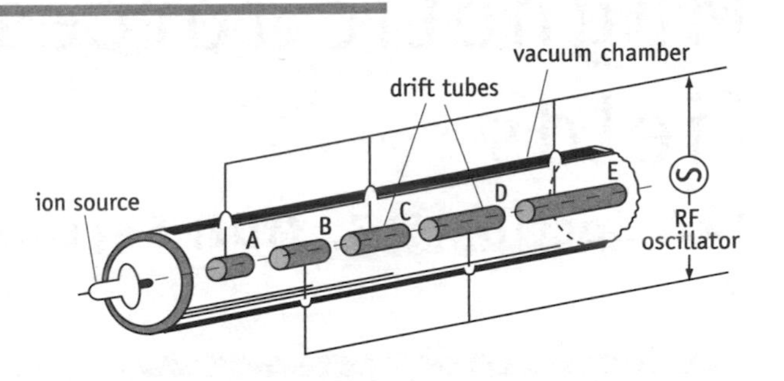

In order to give electrons and other charged particles very high energies, *accelerators* are designed to send the particles through several stages of acceleration.

The diagram shows the principle of a linear accelerator. Suppose the machine shown is using protons.

- The protons will be attracted towards drift tube A when A is negative.
- As the protons pass through A, the alternating voltage will reverse direction, so A will become positive and B will become negative.
- Protons leaving A will be attracted to B. They will experience an acceleration and gain more energy, and so on across each section.
- Because the particles speed up, the drift tubes further down the track are made longer.

Electronvolts

> The eV can be used to express any small amount of energy, not just the energies of electrically accelerated charged particles.

It is often convenient to express energies of subatomic particles in *electronvolts*, eV. 1 eV is the energy transferred to an electron (or proton) when it moves through a potential difference of 1 V.

$$\mathbf{1\ eV = 1.60 \times 10^{-19}\ J}$$

For very high energies, units keV, MeV and GeV are also used:

$$\mathbf{1\ keV = 10^{3}\ eV \quad 1\ MeV = 10^{6}\ eV \quad 1\ GeV = 10^{9}\ eV}$$

To find the energy transferred to a charged particle by an electric field, simply multiply the charge (in units of the electron charge, *e*) by the potential difference.

✓ Quick check 1–3

Worked example

An alpha particle, which has charge +2*e*, is accelerated through a potential difference of 2.5 kV. How much energy does it gain?

$$\mathbf{energy} = 2e \times 2.5 \times 10^{3}\ \mathrm{V} = 5.0 \times 10^{3}\ \mathrm{eV} = 5.0\ \mathrm{keV}$$

Quick check questions

1. A proton is accelerated through a pd of 5.0 kV. What is the energy transferred to it **a** in eV? **b** In J?
2. A gas molecule has kinetic energy 6.21×10^{-21} J. What is this energy in eV?
3. Calculate the voltage required to produce a beam of protons travelling at $1.2 \times 10^{7}\ \mathrm{m\ s^{-1}}$. (Mass of proton = 1.67×10^{-27} kg.)

Magnetic forces and fields

TRA Section 3.3, MDM Section 5.3, PRO Section 4.4

A *magnetic field* is a region in which a magnetic material experiences a force. The strength and direction of a magnetic field can be represented by magnetic field lines. The arrow on a field line shows the direction of the force that would be experienced by the north pole of a magnet placed in that field. Lines closer together represent a stronger field.

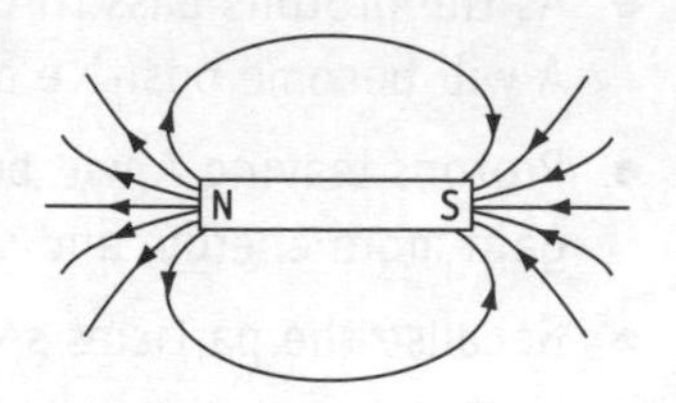

Magnetic field due to a current

An *electric current* has a *magnetic field* around it. If a current flows at an angle to another magnetic field, the two fields interact to produce a force.

The force *always acts at right angles* to both the current and the magnetic field, as shown by Fleming's left-hand rule. Reversing *either* the field *or* the current reverses the direction of the force.

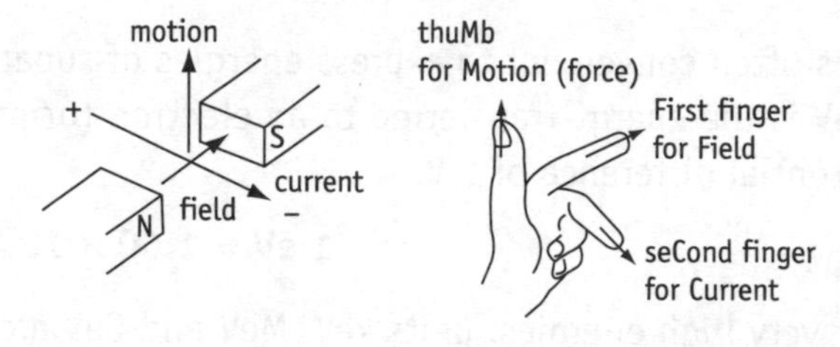

Experiments with a wire and a balance show that the force, F, is directly proportional to:

- the current, I
- the length, ℓ, of wire within the magnetic field
- $\sin \theta$
- the magnetic field strength

$$F \propto BI\ell \sin \theta$$

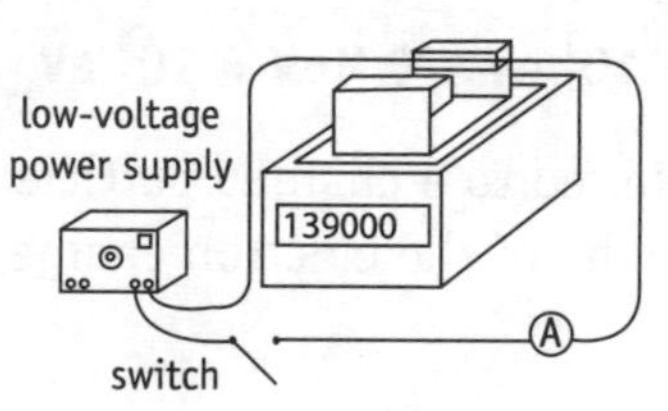

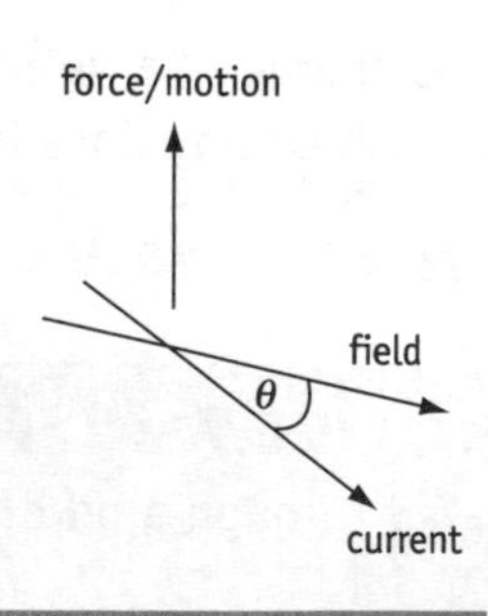

Do not agonise about remembering Fleming's left-hand rule. You will not be expected to use it to predict directions, but it is a useful reminder that the force is always at right angles to the field and current.

Magnetic flux density

Magnetic flux density, B, is a measure of magnetic field strength and is defined as

$$B = \frac{F}{I\ell \sin \theta} \quad \text{so} \quad F = BI\ell \sin \theta$$

The SI unit of magnetic flux density is the tesla, T: 1 T = 1 N A^{-1} m^{-1}.

✓ Quick check 1

Force on a charged particle moving in a magnetic field

When a positive charge q moves with a velocity v at an angle θ to a magnetic field B, it experiences a force at right angles to the field and to its own direction of motion.

$$F = Bqv \sin \theta$$

The direction of the force is reversed by any *one* of the following:

- reverse the field direction
- reverse the direction of the velocity
- change the sign of the charge.

F=Bqv

v

+q

B acting into the page

✓ Quick check 2

The magnetic force on a moving charged particle provides the *centripetal force* necessary for *circular motion*. For a particle mass m moving at 90° to a field B,

$$Bqv = \frac{mv^2}{r}$$

Cancelling v and rearranging gives:

$$r = \frac{mv}{Bq} = \frac{p}{Bq}$$

where p (= mv) is the particle's *momentum*.

✓ Quick check 3, 4

Quick check questions

1 A wire carrying a current of 5.0 A lies in a magnetic field of 80 mT. The length of wire within the field is 2.0 m. Calculate the force on the wire in each of the following cases.

  a The wire is at 90° to the field.

  b The angle between the wire and the field is 60°.

  c The wire lies parallel to the field.

2 a What force acts on an electron moving at 1.0×10^7 m s^{-1} at 90° to a field of 0.20 T?

  b What is the force if the angle between the electron's motion and the field is 40°?

3 A proton (charge +e, mass 1.67×10^{-27} kg) moves at 2.5×10^6 m s^{-1} at right angles to a field of 1.2 T. Calculate the radius of its circular path.

4 In a field of 0.75 T, a particle is observed to travel in a circle of radius 4.5 cm. Assume the particle carries charge e (= 1.60×10^{-19} C) and calculate its momentum. Suppose the particle actually carries charge 2e. How would this affect your answer?

Particles in magnetic and electric fields

MDM Section 5.3, PRO Sections 4.2 and 4.4

The cyclotron

A cyclotron uses a *magnetic field* to bend charged particles into a circular, or spiral, path so that they can be repeatedly accelerated by the same *electric field*.

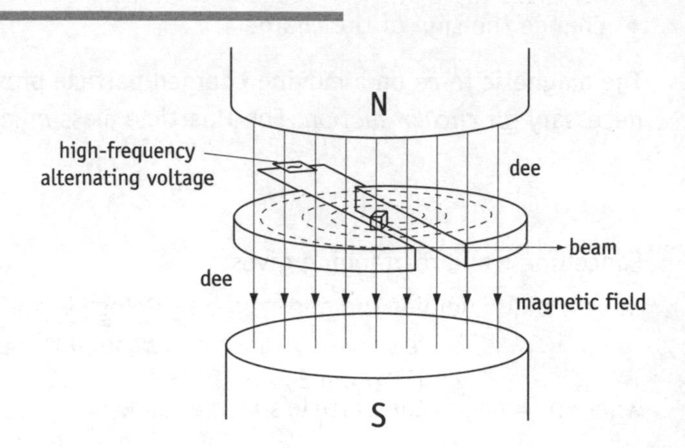

- Protons leaving the source at the centre are attracted towards the negative electrode.
- The magnetic field bends them into a semicircle.
- While they are travelling this semicircle the polarity of the electrodes reverses.
- As they reach the gap, the field between the electrodes acts to accelerate them forwards.

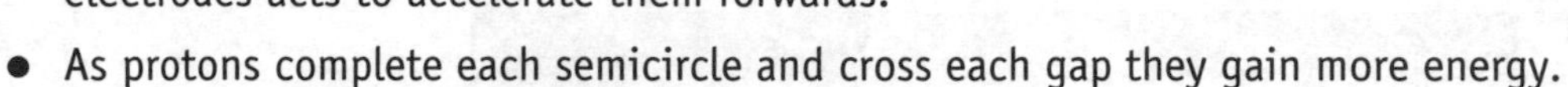

- As protons complete each semicircle and cross each gap they gain more energy.

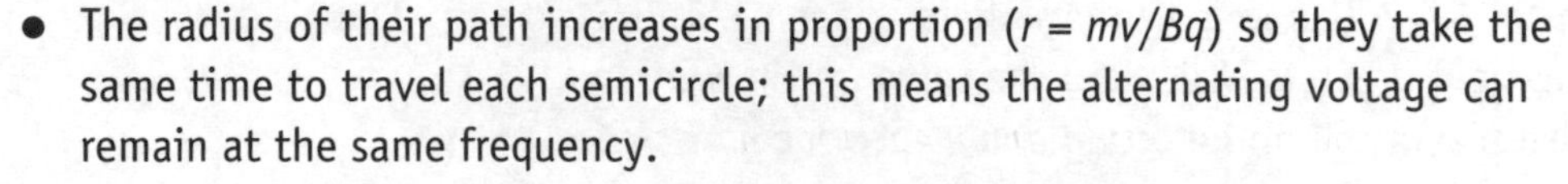

- The radius of their path increases in proportion ($r = mv/Bq$) so they take the same time to travel each semicircle; this means the alternating voltage can remain at the same frequency.
- The protons finally emerge at the outside edge of one of the electrodes and hit their target.

Worked exam question

In a cyclotron protons are accelerated by a high-frequency alternating voltage. A uniform magnetic field, of flux density 200 mT, causes the protons to follow a circular path which increases in radius as the protons gain kinetic energy. Immediately before the protons leave the cyclotron they are moving in a circular arc of radius 1.5 m.

Q Show that the speed of these protons is about 10% of the speed of light. [3]

A Rearrange the equation $r = mv/Bq$ and substitute values:

$$v = \frac{Bqr}{m} = \frac{200 \times 10^{-3}\ \text{T} \times 1.60 \times 10^{-19}\ \text{C} \times 1.5\ \text{m}}{1.67 \times 10^{-27}\ \text{kg}}$$

$$= 2.9 \times 10^{7}\ \text{m s}^{-1}$$

This is close to $c/10$, i.e. 3.0×10^{7} m s^{-1}.

> When asked to 'show that ...', calculate your own answer then compare it with the one given.

Q Calculate the approximate time taken for a proton to complete a semicircle of its orbit. [2]

A Distance round semicircle is πr, and time t is distance/speed:

$$t = \frac{\pi r}{v} = \frac{\pi \times 1.5 \text{ m}}{2.9 \times 10^7 \text{ m s}^{-1}} = 1.64 \times 10^{-7} \text{ s}$$

Q Show that the time is the same for all proton orbits in this field, regardless of speed and orbital radius. [2]

A From above, $t = \pi r/v$. But $v = Bqr/m$, so

$$t = \frac{\pi rm}{Bqr} = \frac{\pi m}{Bq}$$

which does not depend on speed or radius.

Q Calculate the frequency of the accelerating pd. [1]

A Time for complete cycle is $2t$ so

$$f = \frac{1}{2t} = \frac{1}{2 \times 1.64 \times 10^{-7} \text{ s}} = 3.0 \times 10^6 \text{ Hz}$$

The synchrotron

At very high energies the particles travel close to the speed of light, their motion is now described by *relativistic equations* and their travel time is no longer the same for each semicircle. Careful synchronisation is needed to make the electrodes change their polarity as the particles pass through each acceleration section. This type of accelerator is called a *synchrotron*.

Detecting and measuring particles

Electric and magnetic fields also play a role in *detecting and measuring particles*.

- A charged particle moving through matter causes ionisation.
- Early particle experiments used cloud chambers or bubble chambers to show up the trails of ions left behind by particles.
- In multiwire or drift chambers, electric fields accelerate the ions created as the particles pass through.
- The ions are detected and thus the path of the particle can be calculated.
- A magnetic field within the collision region deflects the particles.
- Positive particles bend one way, negative particles bend the opposite way.
- Measuring their curvature allows the physicists to calculate their momentum ($p = BQr$).

✓Quick check 1

? Quick check question

1 For each of the following, say whether it applies to charged particles in **A** electric fields, **B** magnetic fields, or **C** both.

  a The field can transfer energy to the particle.

  b The field can only change the direction of a particle's motion, not its energy.

Electromagnetic induction

TRA Section 3.4

The **magnetic flux** Φ passing at right angles through an area A is given by

$$\Phi = BA$$

where B is the magnetic flux density.

✓ Quick check 1

For a coil of N turns, the **flux linkage** is N times the flux passing through it

$$\text{flux linkage} = N\Phi = NBA$$

The SI unit of flux and flux linkage is the weber (Wb), which is related to the tesla: 1 T = 1 Wb m^{-2}.

✓ Quick check 2

Electromagnetic induction occurs when there is a change in the flux linking a conductor. An emf is induced in the conductor and, if it is part of a complete circuit, an induced current flows. There are three equivalent ways to do this.

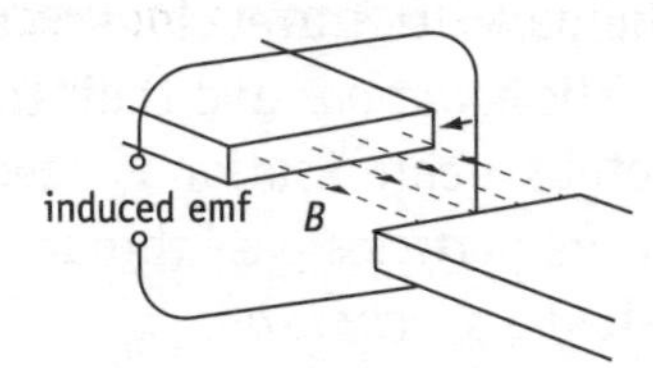

- Move a conductor so that it 'cuts' magnetic flux.
- Move a magnet so that its flux is 'cut'.
- Use a variable electromagnet to change the flux.

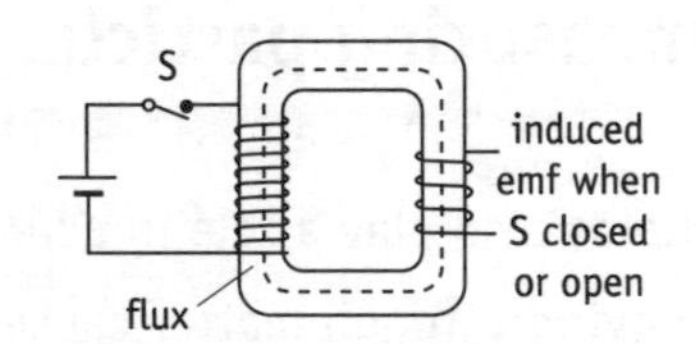

Lenz's law

Lenz's law gives the direction of an induced current or emf. Any forces arising from the induced current act so as to oppose the change causing the induction.

For example, if a straight wire is moved across a magnetic field, current is induced in the wire. This gives rise to a force described by $F = BIl$ and Fleming's left-hand rule. This force opposes the motion of the wire through the field.

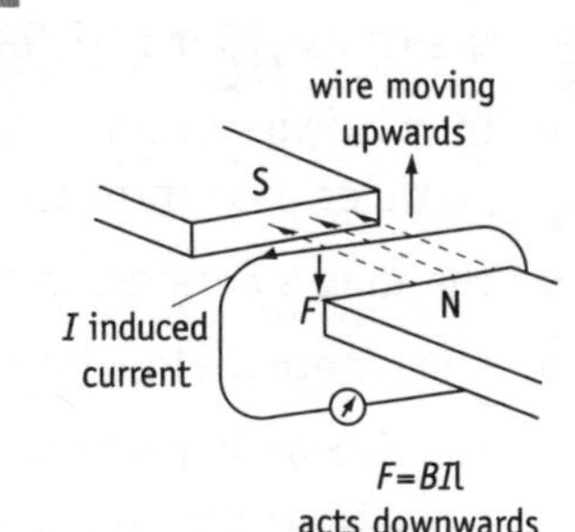

Faraday's law

Faraday's law states that the size of the induced emf, $\mathscr{E}$, is equal to the rate of change of flux linkage:

$$\mathscr{E} = \frac{\textbf{change in flux linkage}}{\textbf{time taken}} = -\frac{d(N\Phi)}{dt}$$

The negative sign is a reminder of Lenz's law.

✓ Quick check 3

Worked examples

1 In the ignition system of a car, a magnetic flux of 7.2×10^{-3} Wb links a coil of 15 000 turns. This flux is reduced to zero in 6.0 ms. What is the induced emf?

Step 1 Calculate the flux linkage:

$$\textbf{flux linkage } N\Phi = 15\,000 \times 7.2 \times 10^{-3} \text{ Wb}$$

Step 2 Use Faraday's law to find the induced emf:

$$\text{induced emf} = \frac{d(N\Phi)}{dt} = \frac{15\,000 \times 7.2 \times 10^{-3} \text{ Wb}}{6.0 \times 10^{-3} \text{ s}} = 1.8 \times 10^{4} \text{ V}$$

2 In a regenerative braking system, the coils of a dynamo are linked to the wheels of the vehicle and an emf is induced. When the braking system is activated, the coils are connected to a battery so that it is recharged by the induced current. Explain **a** why an emf is generated while the vehicle is moving, and **b** why connecting the coils into a circuit causes the vehicle to brake.

a A dynamo contains a *magnet*. If the coil is made to turn within the field (e.g. by the wheels of a moving vehicle), then the *changing magnetic flux* Φ linking the coil gives rise to an *induced emf* as described by Faraday's law $\mathcal{E} = -d(N\Phi)/dt$.

In questions where you are asked to 'explain', try to make your account as precise as possible. Look for opportunities to demonstrate your knowledge of technical terms and quote relevant equations.

b Connecting the moving coil into a complete electric circuit allows a *current* to flow. A wire carrying a current within a magnetic field experiences a force ($F = BI\ell$). The *forces* arising from an induced current always act so as to *oppose* the change that causes the induction, as described by *Lenz's law*. So in this case the forces must oppose the movement of the wheels.

? *Quick check questions*

1 A circular loop of wire with radius 6.0 cm is placed at 90° to a magnetic field of flux density 250 mT. What is the magnetic flux through the loop?

2 The single loop in question **1** is replaced by a coil with the same radius, with 40 turns of wire. What is the flux linkage of the coil?

3 The coil in question **2** is removed from the magnetic field in 0.20 s. What is the size of the emf induced in the coil?

Transformers

TRA Section 3.5

A *transformer* uses a continuously changing magnetic flux to create a continuous induced alternating voltage. The diagram shows the essential features of a transformer: two coils of wire, wound on a common core (usually iron), together with its circuit symbol.

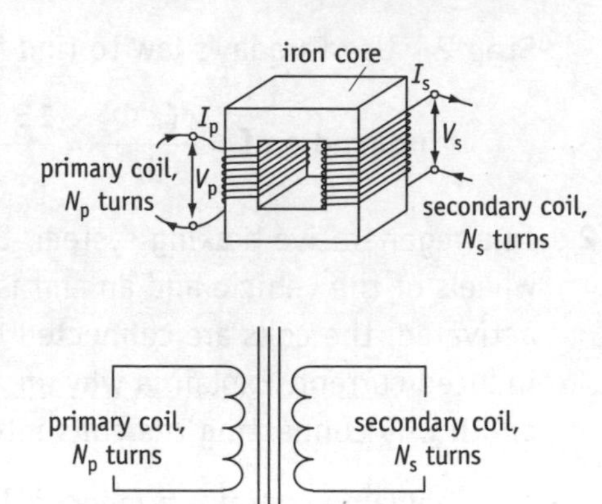

- An alternating voltage is applied to one coil, known as the primary coil.
- This produces an alternating current in the primary coil.
- The alternating current causes changing magnetic flux in the core of *both* coils.
- The continuously changing flux induces an alternating emf in the secondary coil.

The emf in the secondary coil (V_s) is determined by the voltage across the primary (V_p), and the numbers of turns of wire in the secondary and primary coils (N_s and N_p):

$$\frac{V_s}{V_p} = \frac{N_s}{N_p}$$

Thus a transformer with *fewer* turns on the secondary coil than the primary acts as a *step-down* transformer, and one with *more* turns on the secondary acts as a *step-up* transformer.

If the secondary coil is in a complete circuit, then an alternating current flows in it.

The size of the secondary current I_s depends on V_s and on the resistance of the secondary circuit.

Efficiency of a transformer

If a transformer is 100% efficient, then all the power entering the primary also leaves the secondary. Thus

$$\text{power} = V_s I_s = V_p I_p$$

Hence

$$\frac{I_s}{I_p} = \frac{V_p}{V_s} = \frac{N_p}{N_s}$$

Thus the resistance of the secondary circuit also determines the size of the current in the primary coil. A transformer which is being used to step *down* voltage (e.g. a 12 volt lab power supply) will be stepping *up* current in the same ratio.

In practice most transformers are very efficient, though efficiency never reaches 100%.

✓ Quick check 1–3

? Quick check questions

1 A transformer is designed to step a 240 V mains supply down to 24 V.

a What must be the ratio N_s/N_p?

b If the secondary coil is connected to a lamp of resistance 8 Ω, what is the current in the lamp? (Assume the secondary coil has negligible resistance.)

c What is the current in the mains cable?

2 The diagram below shows the alternating primary and secondary voltages in a transformer.

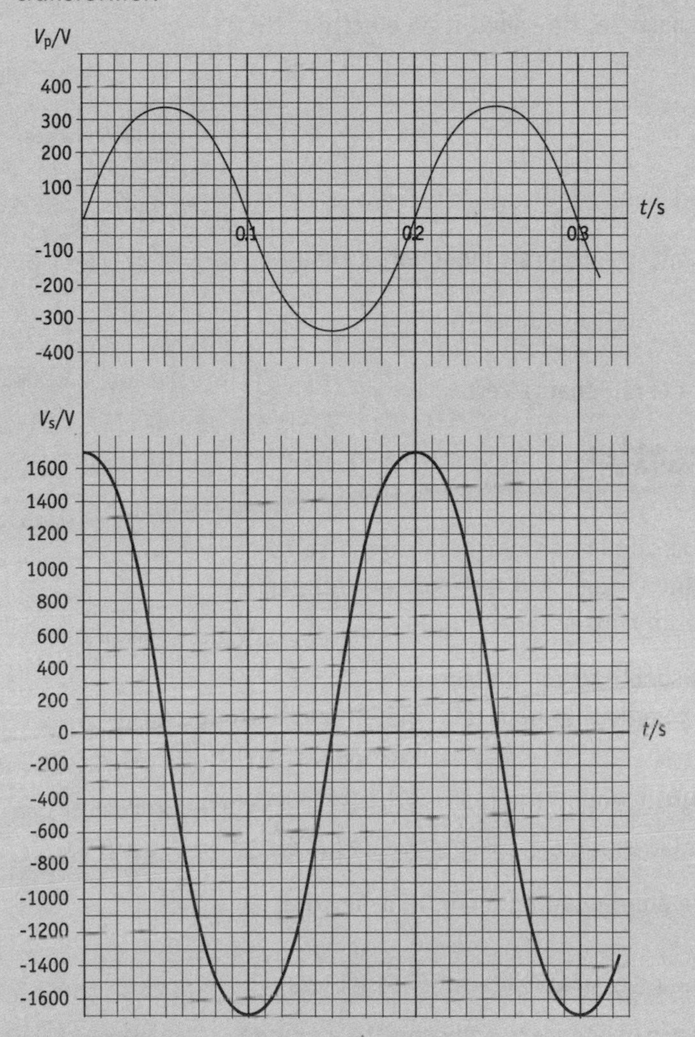

a What is the turns ratio, N_s/N_p, for this transformer?

b Explain why the secondary voltage reaches a peak when the primary voltage passes through zero, and the secondary voltage is zero when the primary voltage is at a peak.

3 Sketch a graph showing how the secondary voltage in the diagram below would vary with time as the switch S is closed, held shut, then opened a few seconds later. Explain the shape of your graph.

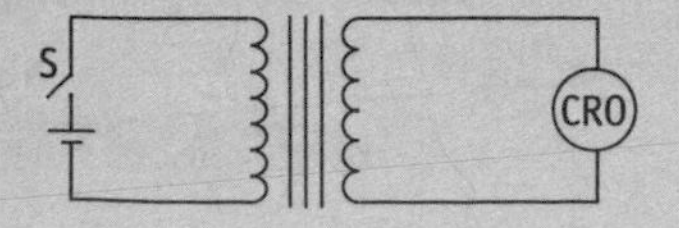

A transformer connected to a battery

Discovering the nucleus

PRO Sections 3.1 and 3.5

For many centuries scientists and thinkers have believed that matter consists of atoms. Until the late 1800s it was thought that atoms were the smallest particles of matter. In 1897 J J Thomson discovered a new particle. He called it an electron. His experiments showed that electrons:

- have a much lower mass than any known atoms
- carry a negative charge
- can be obtained from many different types of atom.

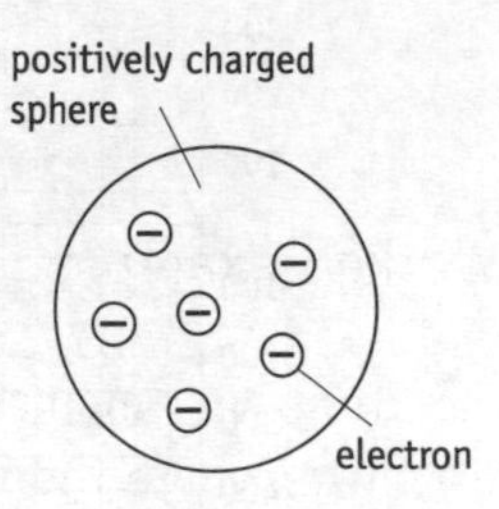

'plum-pudding'- model

Scientists tried to imagine a new model for atoms. Thomson's suggestion, usually called the 'plum-pudding' model, consisted of:

- a sphere of positively charged material
- enough electrons in the sphere to make the overall charge zero.

Using alpha particles to probe matter

Ionising radiations were discovered in the 1890s. Alpha particles were found to be positively charged, with the same mass as a helium atom, and fast-moving. A beam of alpha particles was thus a good tool for probing matter.

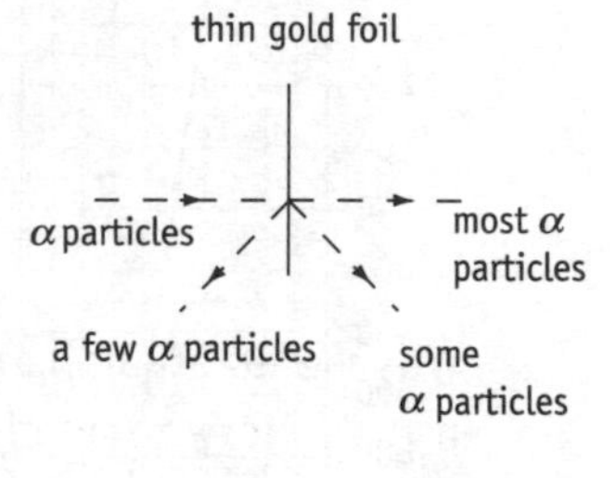

By 1911 Rutherford's co-workers Geiger and Marsden had completed many experiments sending beams of alpha particles towards thin gold foil. They found that:

- nearly all of the alpha particles passed straight through the foil
- a very small fraction were deflected through large angles.

These findings could not be explained using the plum-pudding model. Rutherford proposed a 'nuclear' model.

- Most of the volume of an atom is empty except for electrons.
- There is a tiny nucleus at the centre, with most of the mass and a positive charge that balances the electrons' negative charge.
- If an alpha particle travels close to a nucleus, the electrostatic repulsion and the large mass of the nucleus cause the alpha particle to change direction.
- Most alpha particles do not pass close to a nucleus so they are undeflected.

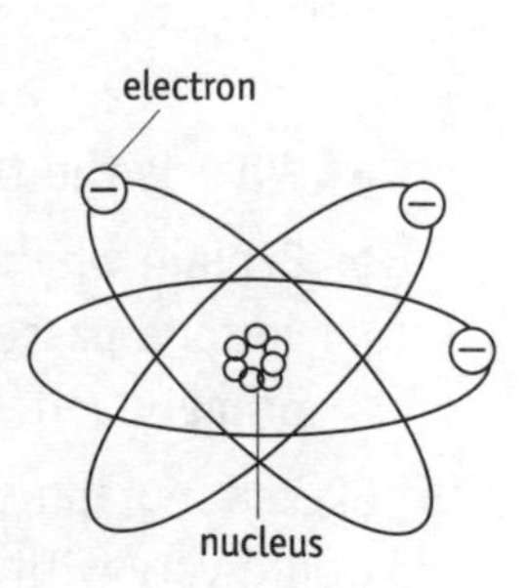

Rutherford's experiments were an early example of an important technique: using a beam of high-energy particles to investigate matter on a subatomic scale. High energies are needed in order to:

- penetrate deep inside composite particles
- disrupt the forces that hold composite particles together.

✓ Quick check 1, 2

Electron diffraction

Another way of thinking about high-energy beams is to deal with their wave properties. Electrons and other particles undergo diffraction, with a wavelength that can be measured using apparatus such as that shown here.

phosphorescent screen
vacuum tube
electron gun
graphite film
diffraction pattern

All particles have an associated *de Broglie wavelength*, λ.

$$\lambda = \frac{h}{p}$$

where p is the particle's momentum and h the Planck constant ($h = 6.63 \times 10^{-34}$ J s).

The smaller the wavelength, the smaller the objects that can be investigated.

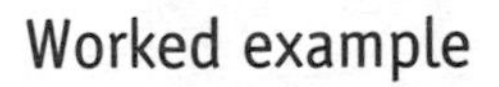

Worked example

Electrons with momentum 5.4×10^{-24} kg m s^{-1} pass through a thin sheet of graphite, where the separation of carbon atoms is approximately 0.37 nm. Explain whether significant diffraction would take place.

Step 1 Find the de Broglie wavelength:

$$\lambda = \frac{h}{p} = \frac{6.63 \times 10^{-34} \text{ J s}}{5.4 \times 10^{-24} \text{ kg m s}^{-1}}$$

$$= 1.2 \times 10^{-10} \text{ m} = 0.12 \text{ nm}$$

As this is about one-third of the atomic separation, there would be significant diffraction.

✓ Quick check 3, 4

? Quick check questions

1. Why would an alpha particle *not* be deflected through a large angle if it passed close to an electron?
2. If atoms were like Thomson's plum-pudding model, what would happen to a beam of alpha particles passing through a thin foil?
3. Electrons accelerated through 200 V are found to have a wavelength of 8.6×10^{-11} m. Calculate their momentum.
4. Suppose the accelerating voltage in question **3** is increased to 800 V. What are the new momentum and wavelength of the electrons?

Particles

PRO Sections 2.2, 2.3 and 2.4

The *standard model* of particle physics describes the basic building blocks of matter. It is the best picture of matter that physicists have at the moment. The standard model identifies twelve *fundamental particles* – that is, particles that cannot be broken into any smaller particles (see table). All matter is made up from combinations of these twelve.

quarks			leptons		
name	**symbol**	**charge**	**name**	**symbol**	**charge**
up	u	$+\frac{2}{3}$	electron	e^-	–1
down	d	$-\frac{1}{3}$	electron-neutrino	ν_e	0
charm	c	$+\frac{2}{3}$	muon	μ^-	–1
strange	s	$-\frac{1}{3}$	muon-neutrino	ν_μ	0
top	t	$+\frac{2}{3}$	tau	τ^-	–1
bottom	b	$-\frac{1}{3}$	tau-neutrino	ν_τ	0

You are not expected to memorise the contents of the table.

For each one of these twelve there is also an *antiparticle*. A particle and its antiparticle have the same mass but opposite charge. The symbol for an antiparticle is generally the same as for the particle, with a bar over the top. For example, the down quark and its antiquark have symbols d and $\bar{d}$ respectively.

Quarks and leptons

The twelve fundamental particles divide into two distinct groups: *quarks* and *leptons*. Within each of these groups there is a clear pattern, with three *generations*. For a number of years the top and bottom quark remained undiscovered. But the *symmetry* of the model convinced physicists that they must exist, and eventually both were found in experiments. Leptons can exist on their own, but quarks exist only in combination with other quarks or antiquarks:

- a quark combined with an antiquark forms a *meson* (e.g. pions, π)
- three quarks combined form a *baryon* (e.g. protons).

pions: π^+ π^0 π^-

e.g. π^- ($\bar{u}$ d)

proton (u u d)

neutron (d u d)

✓ Quick check 1

Everyday matter (atoms and molecules) contains only three of the twelve standard model particles, as shown in the diagram below.

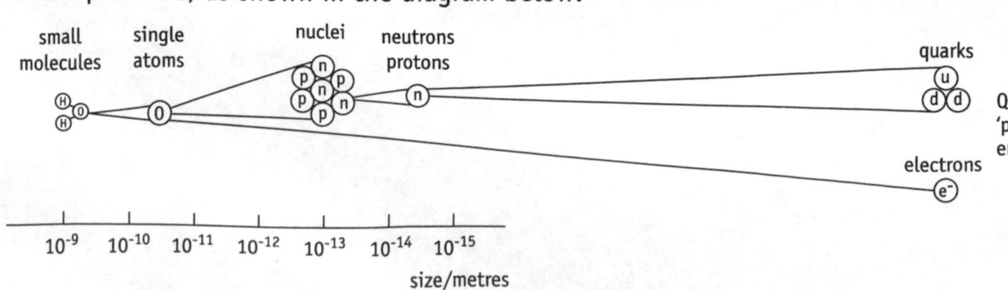

Electron-neutrinos, anti-electron-neutrinos, and anti-electrons (usually called positrons) are emitted during some nuclear reactions. The remaining standard model particles and antiparticles are produced only in accelerator experiments, or in particle reactions in the upper atmosphere or in the hot centres of stars.

All baryons except protons and neutrons, and all mesons, are very unstable, decay quickly, and only occur in accelerator experiments, in the upper atmosphere or in stars.

Any particle smaller than a nucleus is generally known as a *subatomic* particle (though subnuclear might be a better word). Protons and neutrons are collectively known as *nucleons*.

Interactions between particles

Interactions involving subatomic particles and/or nuclei can be represented using symbols.

Positive or negative signs are sometimes written at the top right of particle symbols to indicate their charge in units of e, the charge of a proton ($e = 1.60 \times 10^{-19}$ C).

Nuclei are represented by symbols like this: ${}^{90}_{38}\text{Sr}$

This means the nucleus is the element strontium (Sr), which has 38 protons and 90 nucleons. Therefore we can deduce that it has 90 − 38 = 52 neutrons.

In *all* interactions and decays:

- charge is conserved
- total number of nucleons is conserved.

For example:

$$ {}^{238}_{92}\text{U} \rightarrow {}^{234}_{90}\text{Th} + {}^{4}_{2}\text{He} $$

$$ \text{n} \rightarrow \text{p} + \text{e}^- + \bar{\nu}_e \quad \text{or} \quad {}^{1}_{0}\text{n} \rightarrow {}^{1}_{1}\text{p} + {}^{0}_{-1}\text{e} + \bar{\nu}_e $$

The bottom line of numbers shows charge (92 = 90 + 2; 0 = 1 − 1)

The top line shows nucleon number (238 = 234 + 4; 1 = 1 + 0)

✓ Quick check 2, 3

? Quick check questions

1. A particle is made up of a strange quark s and an up antiquark $\bar{u}$. Is it a meson or a baryon? What is its charge?
2. This equation shows a proton colliding with a lithium nucleus to form a beryllium nucleus. What are the charge and nucleon number of the beryllium?

   ${}^{7}_{3}\text{Li} + \text{p} \rightarrow \text{Be} + \text{n}$
3. Write an equation for a pi-plus (π^+) decaying to an anti-muon and a muon-neutrino.

Particle interactions

PRO Section 2.4

Interactions on a subatomic scale often involve conversions between energy and mass as described by

$$\Delta E = c^2 \Delta m$$

where ΔE is the gain (or loss) of energy, Δm is the loss (or gain) of mass and c is the speed of light ($c = 3.00 \times 10^8$ m s^{-1}).

A particle and its antiparticle may *annihilate* when they meet. Their charges cancel, and their mass turns to energy which appears in the form of electromagnetic photons.

$$e^- + e^+ \rightarrow \textbf{electromagnetic radiation}$$

It is also possible for a particle/antiparticle pair to be created from the energy of a photon. In particle creation and annihilation, charge, momentum and mass/energy are all conserved.

Worked example

An electron and a positron (an anti-electron) annihilate forming two identical photons. What is the wavelength λ of each photon? (Electron mass $m_e = 9.11 \times 10^{-31}$ kg.)

Step 1 Find the change in mass:

$$\Delta m = 2 \times m_e$$

Step 2 Use the conversion of mass to energy:

$$\text{energy of photons} = c^2 \Delta m = 2 m_e c^2$$

so each photon has energy $E = m_e c^2$.

Step 3 Use the relationship between energy and frequency:

$$E = hf \quad \text{and} \quad c = f\lambda \quad \text{so} \quad E = \frac{hc}{\lambda}$$

$$\text{hence } \lambda = \frac{hc}{E} = \frac{hc}{m_e c^2} = \frac{h}{m_e c}$$

$$= \frac{6.63 \times 10^{-34} \text{ J s}}{9.11 \times 10^{-31} \text{ kg} \times 3.00 \times 10^8 \text{ m s}^{-1}} = 2.43 \times 10^{-12} \text{ m}$$

You could calculate E then find f and hence λ. But it is much better to do the algebra first as it shows how to get straight to the final answer.

✓ Quick check 1

Particle physicists sometimes express mass in non-SI units MeV/c^2 and GeV/c^2, because a mass in these units can easily be related to the equivalent amount of energy.

Worked example

A proton has mass $m_p = 1.67 \times 10^{-27}$ kg. Calculate its mass in MeV/c^2.

Step 1 Calculate the energy equivalent to this mass in joules.

$$E = c^2 m_p = (3.00 \times 10^8 \text{ m s}^{-1})^2 \times 1.67 \times 10^{-27} \text{ kg}$$

$$= 1.50 \times 10^{-10} \text{ J}$$

Step 2 Convert this energy to MeV:

$$1 \text{ MeV} = 1.60 \times 10^6 \times 10^{-19} \text{ J} = 1.60 \times 10^{-13} \text{ J}$$

so

$$1 \text{ J} = \frac{1 \text{ MeV}}{1.60 \times 10^{-13}} = 6.25 \times 10^{12} \text{ MeV}$$

so

$$c^2 m_p = 1.50 \times 10^{-10} \text{ J} \times (6.25 \times 10^{12} \text{ MeV J}^{-1}) = 938 \text{ MeV}$$

Step 3 Divide both sides of the equation by c^2:

$$c^2 m_p = 938 \text{ MeV} \quad \text{so} \quad m_p = 938 \text{ MeV}/c^2$$

✓ Quick check 2–4

? *Quick check questions*

1. A proton annihilates with an anti-proton to produce two photons. Calculate the wavelength of each photon. ($m_p = 1.67 \times 10^{-27}$ kg.)
2. The mass of a top quark is 180 GeV/c^2. What is this mass in kilograms?
3. The charm quark has mass 1.3 GeV/c^2 and charge $+\frac{2}{3}e$. What are the mass and charge of its antiparticle?
4. A bottom quark annihilates with its antiquark, producing two identical photons. The mass of a bottom quark is 4.3 GeV/c^2. Calculate the wavelength of each photon.

Practice exam questions for PSA4

1 This question is about modelling a collision interaction between two objects.

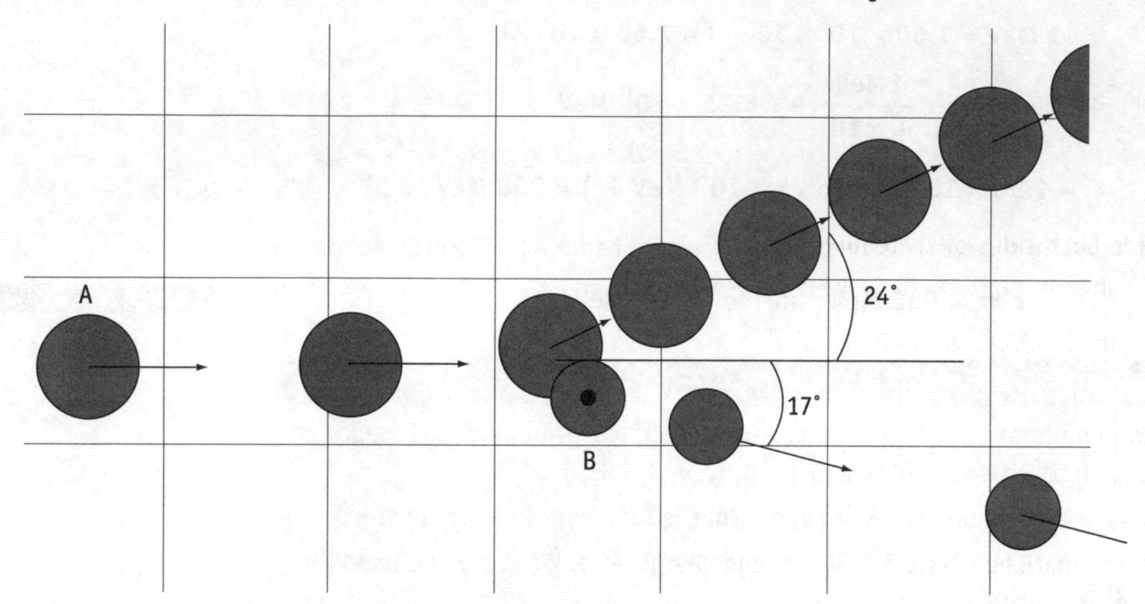

The diagram is a computer-generated model of a two-dimensional collision and is drawn to scale. It shows the positions of two objects at successive equal time intervals. The large object A has mass 10 kg, and initially moves at 5.0 m s^{-1}. Object B has mass 5.0 kg, and is initially stationary.

Take measurements of the motion of A before the collision and use them to calculate the scale of the diagram.

Hence show that after the collision the speed of object A is about 2.3 m s^{-1}. [2]

Make a reasoned estimate of the percentage uncertainty in the value of this speed, as measured from the diagram. [2]

The speed of object B after the collision, as measured from the diagram, is 6.0 m s^{-1}. Make calculations to decide whether the collision is elastic or inelastic. Show your reasoning and conclusion clearly. Consider how the uncertainties in the measurements may affect your conclusion. [4]

After the collision, object A has a component of momentum in the $+y$ (up the page) direction, and object B has a component in the $-y$ direction. Explain why you would expect the values of these two components to be equal. [1]

Calculate these two components of momentum and comment on your answers. [3]

From Edexcel June 2000, Unit PSA4 [Total: 12]

2 Transmitting speech by telephone usually involves transmitting frequencies up to about 3500 Hz. In a digital system this requires a sampling rate of around 8 kHz.

A good quality telephone connection can be achieved with 64 000 bits (pulses) per second. We can now generate over 2 billion (2×10^9) pulses per second along a single fibre, enough for a very large number of telephone connections.

A single fibre in this system is capable of handling about 32 000 telephone calls simultaneously.

Explain briefly why 'transmitting speech by telephone usually involves transmitting frequencies up to about 3500 Hz'. [1]

Explain why a sampling rate of around 8 kHz is used in a digital system. [2]

Using the data above, calculate how many bits are used to encode each sample. [2]

Show that 'a single fibre in this system is capable of handling 32 000 telephone calls simultaneously'. State how this is achieved. [2]

From Edexcel January 2000, Unit PSA4 [Total: 7]

3 The Large Electron Positron collider or LEP, which is at CERN in Switzerland, is the largest machine of its type in the world. It is circular, with a circumference of 27 km, and is used to accelerate electrons and positrons to very high energies. These high-energy particles are then made to collide with each other to produce rare particles for the scientists to observe.

Calculate the radius of the LEP. [1]

The electrons entering the LEP ring already have kinetic energy of about 20 GeV. Explain what this means. [1]

At this energy the speed of the electrons is very close to the speed of light in a vacuum and their mass is about 3.6×10^{26} kg. Comment briefly on this mass. [1]

Calculate the centripetal acceleration of one of these electrons travelling around the LEP. Assume that the speed of the electron is 3.0×10^8 m s^{-1} to two significant figures. [2]

Show that the centripetal force needed to keep an electron in this orbit is about 8×10^{-13} N. [1]

Calculate the magnetic flux density needed to keep an electron in this orbit. [2]

Whilst the electrons are in the LEP, their energy is increased to 600 GeV. The magnetic flux density required to keep them in a constant orbit has to be increased. Explain why this is necessary. [2]

From Edexcel June 2000, Unit PSA4 [Total: 10]

4 Most types of microphone detect sound because the sound waves cause a diaphragm to vibrate. In one type of microphone this diaphragm forms one plate of a parallel plate capacitor. As the diaphragm moves, the capacitance changes. Moving the plates closer together increases the capacitance. Moving the plates further apart reduces the capacitance.

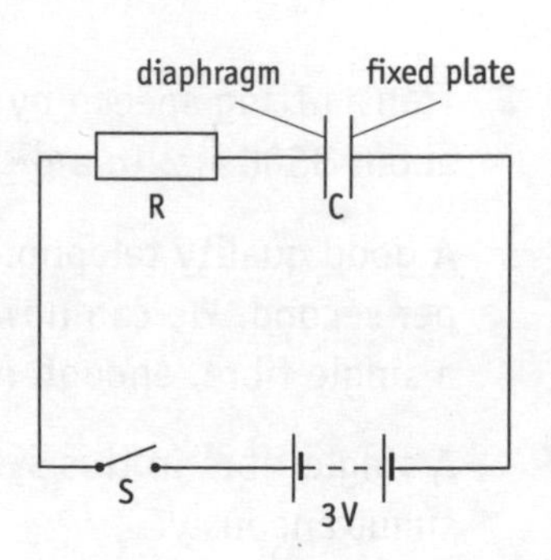

This effect is used to produce the electrical signal. The circuit shown here consists of a 3 V supply, an **uncharged** capacitor microphone C, a resistor R, and a switch S.

The switch S is closed. Sketch a graph of the voltage across the capacitor microphone against time. Assume that the capacitor microphone is not detecting any sound. [3]

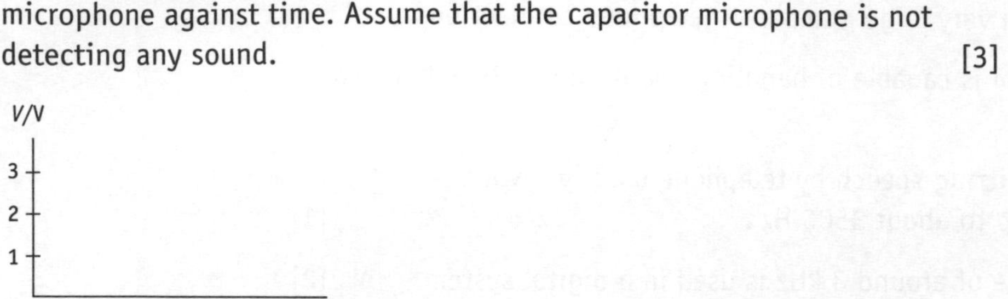

Explain why movement of the diaphragm causes a potential difference (the signal) across R. [4]

From Edexcel June 2001, Unit PSA4 [Total: 7]

5 The ignition system in a car requires 25 000 V to be applied to the spark plug to produce a spark in the combustion chamber. This voltage is produced from the car's 12 V dc electric supply by using a type of transformer usually called the 'ignition coil'. A circuit diagram of such a coil is shown to the right.

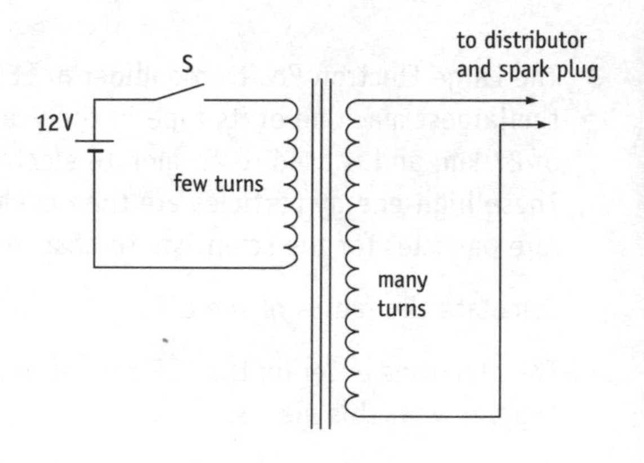

In order to generate a pulse of high voltage at the spark plug, the switch S must be closed for a short period and then opened quickly.

Use Faraday's law to explain why a large voltage is generated in the secondary circuit when the switch is opened.

From Edexcel January 2001, Unit PSA4 [Total: 6]

6 Consider the electric field near a single proton. Calculate the distance r from the single proton at which the field has a value 5×10^{10} V m^{-1}. [3]

Using your knowledge about atoms, comment on your answer. [1]

From Edexcel June 2001, Unit PSA4 [Total: 4]

7 The results of particle collisions are observed using particle detectors. Sometimes the products of these collisions are themselves unstable and decay to give further particles; the products of decay can be photons.

In one collision, a stationary π° meson is produced which then decays to give two gamma ray photons. The rest mass of the meson is 135 MeV/c^2.

$$\pi^{\circ} \rightarrow \gamma + \gamma$$

Calculate the wavelength of each photon. [5]

From Edexcel June 2001, Unit PSA4 [Total: 5]

Unit PSA5: Physics from Creation to Collapse

This unit tests:

- **Build or Bust? (BLD)**
- **Reach for the Stars (STA)**

This part of the revision guide follows the same order as your textbook. The heading of each spread indicates the physics content covered and there is a reference to the relevant sections(s) of your textbook where you will find further details.

Simple harmonic motion

BLD Section 2.2

Simple harmonic motion (SHM) occurs when the resultant force, F, acting on an object is proportional to its displacement, x, from equilibrium and in the opposite direction

$$F = -kx$$

where k is a constant (sometimes called the *force constant*). A good example of SHM is a mass held between two springs and moving back and forth. If the springs obey Hooke's law, they give a force proportional to displacement.

Analysing SHM

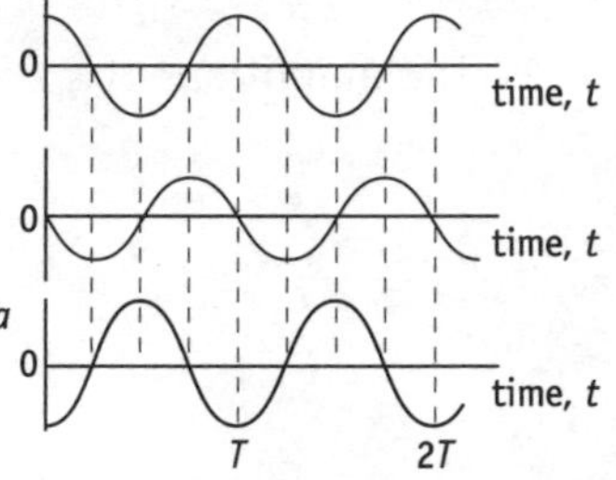

The displacement of a simple harmonic oscillator varies with time in a sinusoidal way. If the displacement is maximum at $t = 0$, then

$$x = A \cos(\omega t)$$

where A is the *amplitude* and ω the *angular frequency*, normally expressed in radians per second.

Angular frequency ω is related to the *period*, T, of the oscillations:

$$\omega = \frac{2\pi}{T}$$

When doing calculations with cos ω or sin (ωt), remember to switch your calculator into radian mode.

The velocity, v, can be found from the gradient of the displacement–time graph. This graph is described by

$$v = -A\omega \sin(\omega t)$$

The acceleration, a, is found from the gradient of the velocity–time graph and is described by

$$a = -A\omega^2 \cos(\omega t)$$

Note the following:

- The maximum possible value of sin (ωt) or cos (ωt) is 1.
- The maximum displacement is A.
- The maximum velocity is equal to $A\omega$.
- The maximum acceleration is equal to $A\omega^2$.

By comparing the expressions for displacement and acceleration, we can deduce that

$$a = -\omega^2 x$$

Since $a = F/m$ where m is the object's mass, and $F = -kx$, we have

$$\omega^2 = \frac{k}{m}$$

✓Quick check 1–3

Worked example

A mass held between two springs is oscillating with an amplitude of 0.10 m and a period of 2.0 s.

a Calculate the angular frequency ω.

b The motion begins with the mass at maximum displacement when $t = 0$. Calculate the displacement 2.7 s later.

a Use the relationship between angular frequency and period:

$$\omega = \frac{2\pi}{T} = \frac{2\pi}{2\text{ s}} = \pi \text{ rad s}^{-1} = 3.14 \text{ rad s}^{-1}$$

b Use $x = A \cos(\omega t)$

$$\omega t = 3.14 \text{ rad s}^{-1} \times 2.7 \text{ s} = 8.48 \text{ rad}$$

$$x = 0.10 \text{ m} \times \cos(8.48 \text{ rad}) = 0.10 \text{ m} \times -0.59 = -0.059 \text{ m}$$

Take care with signs. A negative displacement means that the object is displaced in the opposite direction to its initial displacement.

Quick check questions

1 The amplitude of an oscillator is 0.20 m and its period is 3.0 s. Calculate
  a the angular frequency ω
  b the maximum speed
  c the maximum acceleration.

2 An oscillator has angular frequency 7.4 rad s^{-1} and amplitude 2.5×10^{-3} m. It has its maximum displacement when $t = 0$. Calculate its displacement, velocity and acceleration when $t = 2.0$ s.

3 Explain whether any of these is an example of simple harmonic motion.
  a A person jumping up and down on the ground.
  b A baby bouncing gently in an elastic harness.
  c A person trampolining.

Energy, damping and forced oscillations

BLD Sections 2.1, 2.2 and 3.1

The *kinetic energy*, E_k, of a simple harmonic oscillator is continuously changing.

- E_k is maximum when speed is maximum. This occurs when the displacement $x = 0$ (the equilibrium position).
- The kinetic energy is zero when displacement is maximum (the oscillator is momentarily at rest).

$$E = \frac{1}{2} mv^2 = \frac{1}{2} m\,(A\omega \sin(\omega t))^2 = \frac{1}{2} mA^2\omega^2 \sin^2(\omega t)$$

The *potential energy*, E_p, is also changing continuously.

- E_p is maximum when the displacement is greatest.
- E_p is zero as the oscillator moves through its equilibrium position.

$$E_p = \frac{1}{2} kx^2 = \frac{1}{2} kA^2 \cos^2(\omega t) = \frac{1}{2} mA^2\omega^2 \cos^2(\omega t)$$

If no energy is lost from the oscillator, the sum of its kinetic and potential energy remains constant throughout its motion.

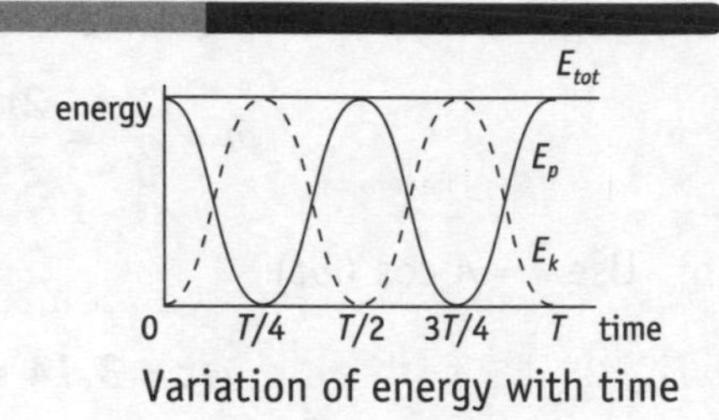

Variation of energy with time

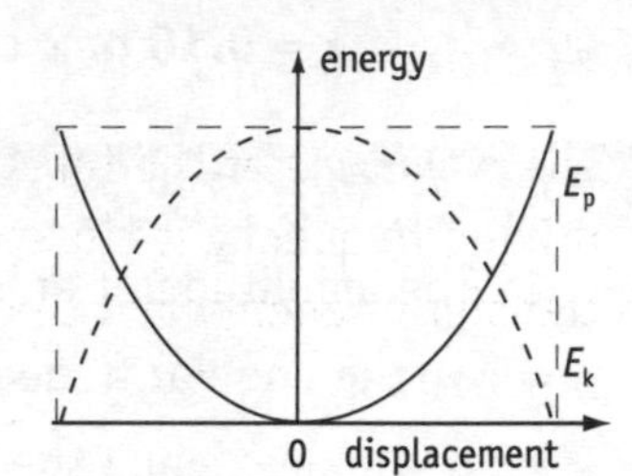

Variation of energy with displacement

✓ Quick check 1, 2

Types of oscillation

A system that oscillates with no input or loss of energy is said to perform **free oscillations**. The frequency of these oscillations depends only on the physical properties of the system (e.g. mass and force constant). This is the system's **natural frequency** of oscillation.

If energy is lost, then the motion is said to be **damped**. The amplitude decreases *exponentially* with time as shown in the figure. Such energy losses can be produced by friction and air resistance, which lead to heating of the oscillator and its surroundings.

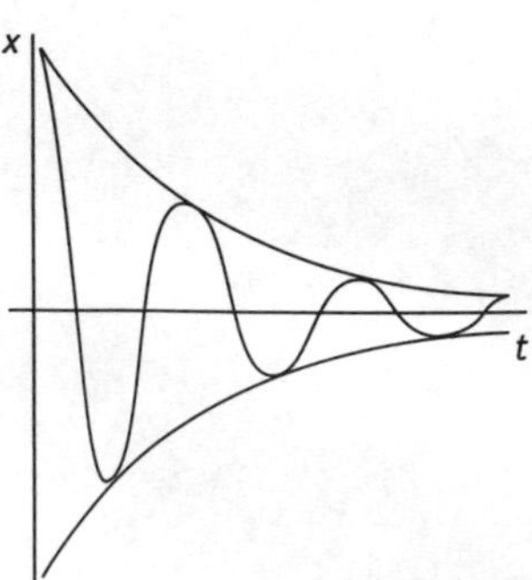

▸▸ *Exponential changes are discussed in PSA4.*

An oscillator can be made to undergo **forced oscillations.** An external force is applied by another oscillating system, and energy is transferred to the original oscillator. In the figure a vibration generator makes a mass on a spring perform forced oscillations.

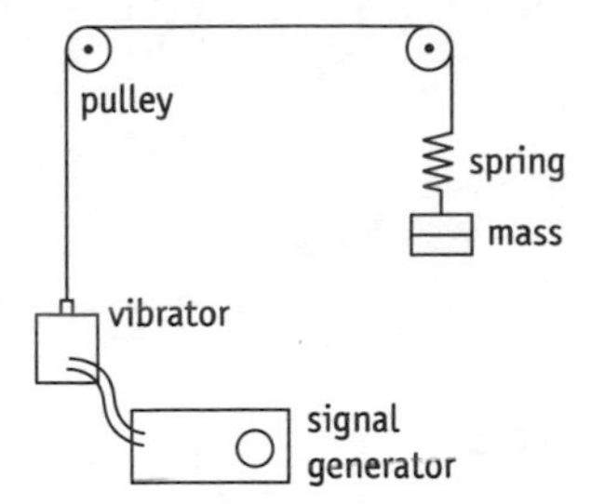

The *frequency* of the forced oscillations is the same as the frequency of the system used to force them.

Resonance

The *amplitude* of the forced oscillations depends on the driving frequency. When the driving frequency is close to the oscillator's own natural frequency, there is a large transfer of energy and the oscillations build up to large amplitude. This is known as **resonance.**

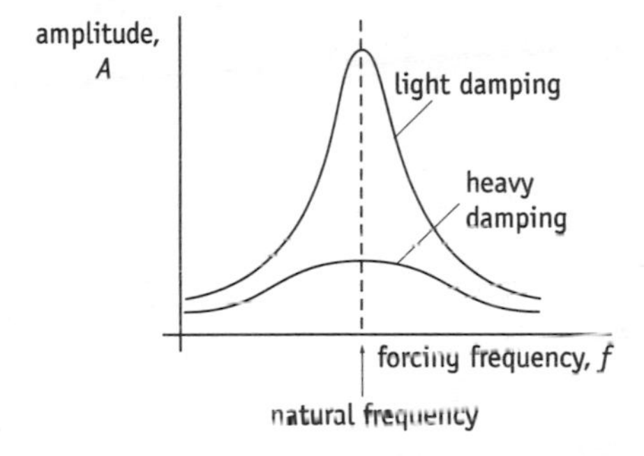

Damping reduces the amplitude of the forced oscillations.

✓ Quick check 3

Worked exam question

A student investigated the behaviour of an oscillating 100 g mass on a spring using a data logger to produce the graph shown.

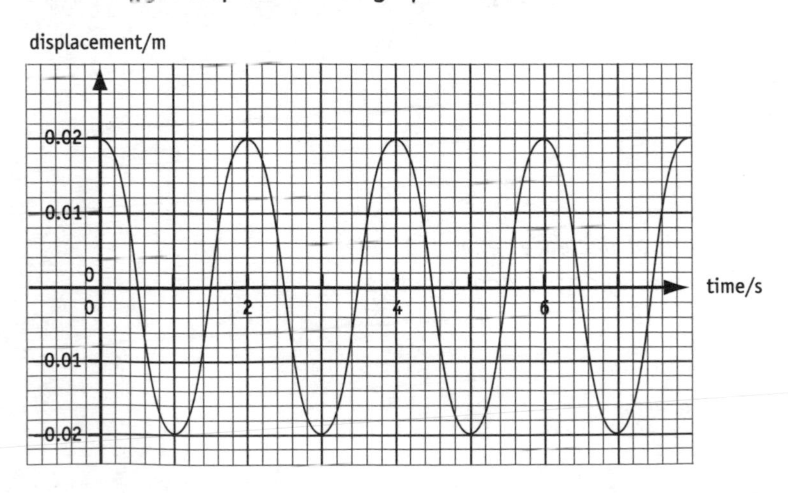

The equation specifying the variation of its acceleration with time is given as:

$$a = -A\omega^2 \cos(\omega t)$$

Q Use the graph above to obtain values for A and ω. [3]

A A **= 0.02 m** (maximum displacement, read directly from graph)

$$T = 2.0 \text{ s} \quad \text{so} \quad \omega = \frac{2\pi}{T} = \frac{2\pi}{2.0 \text{ s}}$$

$$\omega = \pi \text{ rad s}^{-1} = 3.14 \text{ rad s}^{-1}$$

Q Calculate the maximum kinetic energy of the 100 g mass. [3]

A Max E_k = total energy

$$\text{total energy} = \tfrac{1}{2} mA^2\omega^2 = \tfrac{1}{2} \times 0.100 \text{ kg} \times (0.02 \text{ m} \times 3.14 \text{ rad s}^{-1})^2$$

$$= 2.0 \times 10^{-4} \text{ J}$$

Alternatively, use $v = A\omega = 0.063$ m s^{-1}, *and then* $E_k = \tfrac{1}{2}mv^2$.

Remember to convert mass to kg.

Q The mass on the spring was then suspended in a beaker of water and oscillated, and a second graph was plotted. Use this graph, shown below, to describe the behaviour of the mass in the water. [4]

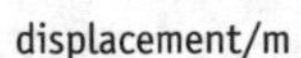

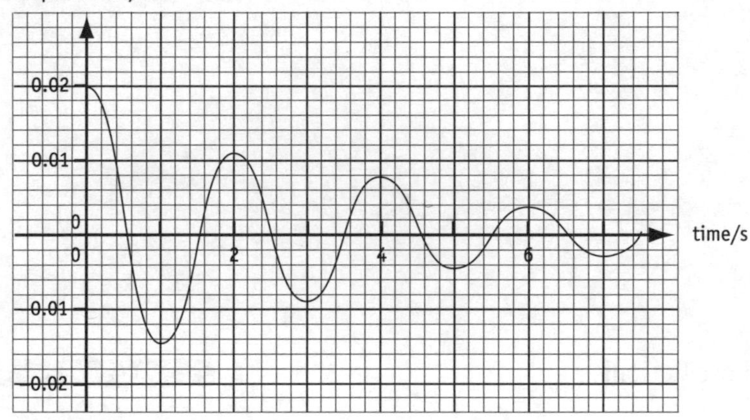

A You would get one mark each for any four of the following:

- Resistive forces in the water
- dissipate energy from the oscillations
- and so the oscillations are damped.
- The time period remains constant.
- The amplitude decreases
- exponentially with time.

Be careful with your use of words. It is the amplitude *that decreases exponentially, not the* displacement. *The displacement oscillates between positive and negative values and passes through zero several times.*

? Quick check questions

1 Write down expressions for the following:

a The maximum kinetic energy of an oscillator in terms of m, A and ω.

b The maximum potential energy of an oscillator in terms of k and A.

2 A mass 0.10 kg undergoes SHM with amplitude 0.20 m and period 3.0 s. Calculate:

a its maximum kinetic energy;

b the force constant k; and

c the displacement when the kinetic and potential energies are equal.

3 A car-driver notices that, when she drives at a particular speed, the rear-view mirror shakes noticeably. In an attempt to stop it, she sticks a large lump of Blu-tack to the back of the mirror.

a Explain why the vibration is noticeable only at a particular speed;

b Explain how the Blu-tack will affect the vibrations.

Seismic and sound waves

BLD Sections 1.2, 2.3, 3.2 and Part 4

Sound waves and seismic P-waves are both examples of longitudinal waves. Such waves travel through solids with a speed, v, given by

$$v = \left(\frac{E}{\rho}\right)^{\frac{1}{2}}$$

where E is the *Young modulus* of the material and ρ its density.

✓ Quick check 1

The **intensity** of a sound is the rate at which energy crosses unit area. The SI units of intensity are W m^{-2}. Intensity is proportional to the *square* of the wave amplitude.

✓ Quick check 2

Absorption of wave energy

If sound or seismic waves travel through material that deforms plastically (i.e. a *ductile* material), the oscillations are damped and the wave amplitude is reduced. Damping by absorbent materials can be used to reduce the damage caused by earthquakes and the nuisance caused by unwanted sound (noise).

Sound absorption can be increased if the material is **porous**, i.e. is made up of small cells or fibres with air trapped between them. The sound waves make the air pockets vibrate, the sound is reflected at the boundaries rather than travelling onwards, and plastic deformation of the cells and fibres damps the vibrations and reduces the amplitude of the sound wave.

Active noise control

If materials alone cannot absorb sound which is a nuisance, then **active noise control** can be used.

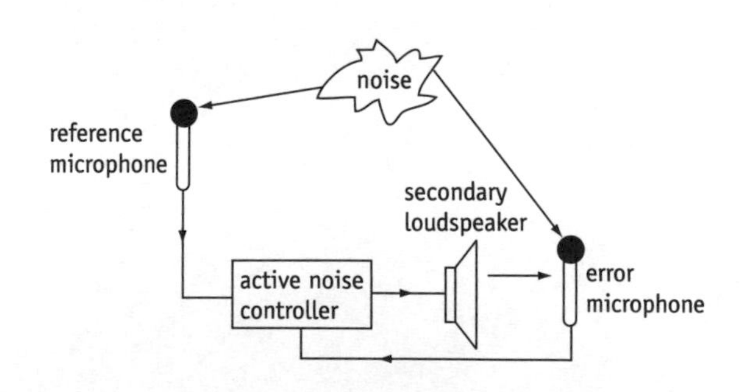

- A reference microphone detects the noise.

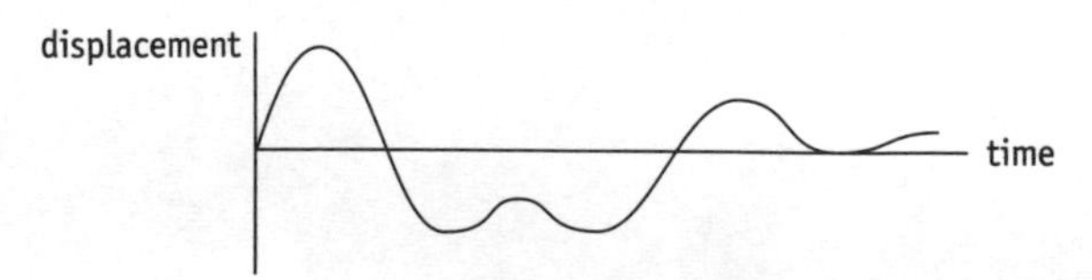

- This signal is then amplified.
- It is then inverted to provide **negative feedback,** i.e. it is half a cycle (180°, or π rad) out of phase with the original.

- This signal drives a speaker.

Inverted signal from speaker

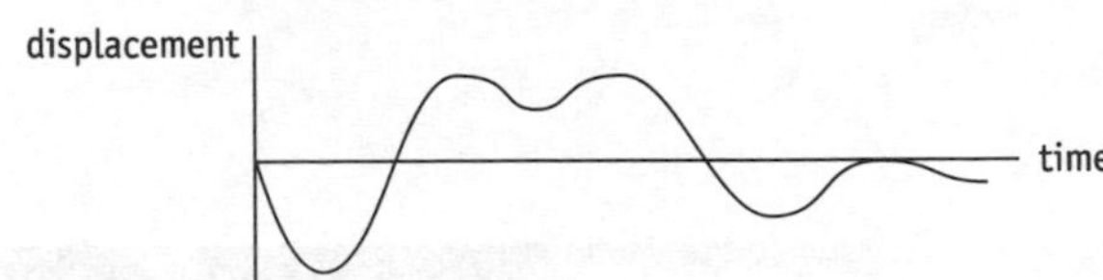

- The sound from the speaker undergoes **superposition** with the original sound so that the resulting amplitude is reduced.
- An error microphone detects the resulting sound and sends a signal to adjust the amplification.
- If the correct amplification is used, and the system can respond rapidly enough to changes in noise level, then the resultant amplitude is zero.

Resultant signal at error microphone

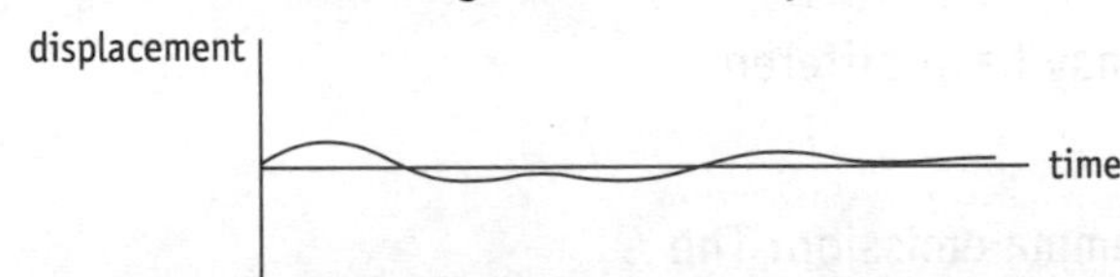

? *Quick check questions*

1. The speed of sound in steel is 5100 m s^{-1} and its density is 7700 kg m^{-3}. What is its Young modulus?
2. An absorbing material successfully reduces the amplitude of some unwanted sound to one half its initial value. By what fraction will the intensity of this sound be reduced?

Practise summarising the key features of active noise control and sketching diagrams similar to the figures here.

Radioactive decay

STA Section 2.2

Nuclei are represented by symbols like this: $^{235}_{92}U$

This represents the element uranium (U). Here the **proton number** is 92 and the **nucleon number** is 235.

Nucleon number is also called the **mass number**. Each nucleon (proton or neutron) has mass 1.67×10^{-27} kg.

The proton number indicates the nuclear charge, as each proton has charge $e = 1.60 \times 10^{-19}$ C.

Isotopes of an element all have the same proton number but may have different numbers of neutrons and hence different mass numbers.

Radioactive isotopes are nuclei that decay by alpha, beta or gamma emission. The resulting nucleus is known as the *daughter*.

For example

$$^{220}_{86}\mathbf{Rn} \rightarrow {}^{216}_{84}\mathbf{Th} + {}^{4}_{2}\mathbf{He}\ (\alpha)$$

The proton numbers and mass numbers balance on both sides of the equation. The bottom line of numbers shows charge (86 = 84 + 2). The top line shows nucleon number (220 = 216 + 4).

✓ Quick check 1

Decay constant and sample activity

In a sample of radioactive material, it is impossible to predict which nucleus will decay next or when a particular nucleus will decay. But if there is a large enough number of radioactive nuclei it is possible to predict the fraction that will decay in a given time interval. The fraction that decays per unit time is known as the **decay constant**, λ.

The **activity**, A, of a sample is normally measured as the number of nuclei decaying per second. Activity is proportional to the number, N, of nuclei present

$$A = \frac{dN}{dt} = -\lambda N$$

This decay equation can be written in another way to show that number and activity decrease *exponentially* with time.

$$N = N_0\, e^{-\lambda t} \quad \text{or} \quad A = A_0\, e^{-\lambda t}$$

where N_0 is the number of nuclei and A_0 the activity at $t = 0$.

The SI unit of activity is the becquerel, Bq. 1 Bq = 1 decay per second = 1 s^{-1}. For an activity to be expressed in Bq, the decay constant must be expressed as the fraction decaying *per second*.

Half-life

The *half-life*, $t_{1/2}$, is the time taken for half of a particular radioactive isotope to decay. If $N = N_0/2$ then

$$\frac{1}{2} = e^{-\lambda t_{1/2}}$$

Taking reciprocals then natural logs (ln) of both sides

$$\ln(2) = \lambda t_{1/2}$$

Half-life can be found experimentally by making several measurements of the time for the activity of an isotope to halve then averaging the result.

✓ Quick check 2

Worked example

The figure shows how the activity of a radon sample (^{220}Rn) changes with time. By making two estimates of half-life from the graph, determine the half-life of ^{220}Rn.

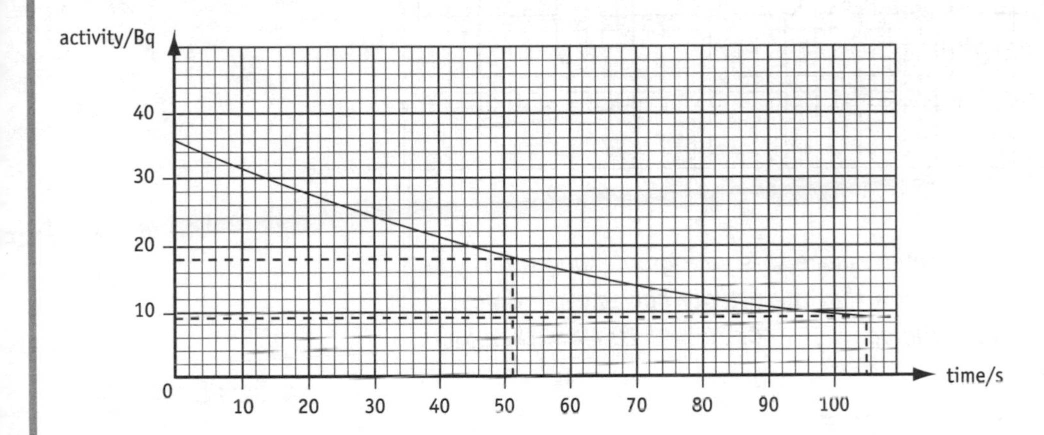

When $t = 0$, $A = 36$ Bq; when $A = 18$ Bq, $t = 52$ s.

For a second measurement, we can use the time taken for A to halve again to 9 Bq. When $A = 9$ Bq, $t = 105$ s so time taken = 53 s.

Averaging the two time intervals give $t_{1/2} = 52.5$ s.

Any two convenient starting points can be used and all will give very similar answers. For example, you could find the time for the activity to decay from 30 Bq to 15 Bq, or from 20 Bq to 10 Bq.

Worked exam question

The 'Radiocat' website says:

> Older cats often get benign tumours of the thyroid gland, but radiocat can deal with these. One injection of radioiodine $^{131}_{53}I$ is all it takes! The $^{131}_{53}I$ is absorbed into the thyroid and destroys the tumour. Our treatment plan includes $^{131}_{53}I$ injection and daily monitoring with as much love and care as we can safely give. Your cat must reach a safe and legal level of radioactivity before coming home.

Q The $^{131}_{53}I$ has a half-life of 8 days and, when first injected, its activity is 80 MBq. On the axes below, sketch a graph showing how the activity changes with time for 24 days after the injection. [2]

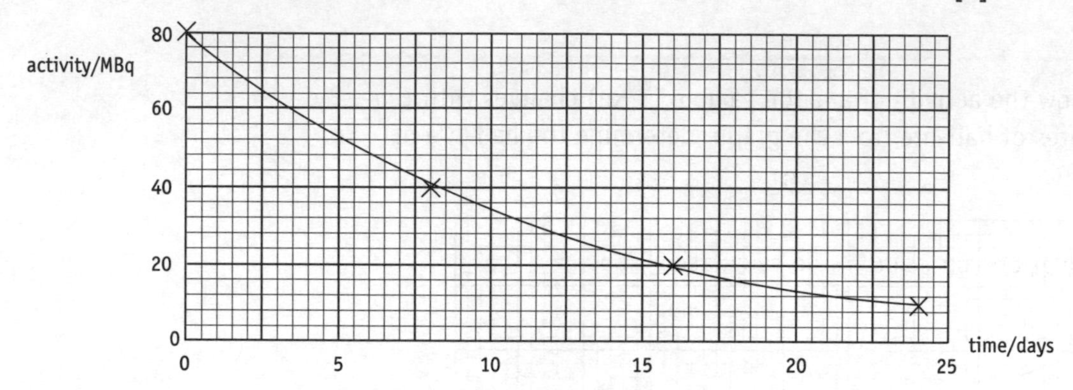

Even though you are asked for a 'sketch', make sure your graph shows that you know the relevant physics. It must show that the activity halves every 8 days. Plot points every 8 days to show when the activity is 40, 20 and 10 MBq, then join them with a smooth curve.

Q Describe the composition of the nucleus $^{131}_{53}I$. [2]

A It contains 53 protons and 131 − 53 = 78 neutrons.

Q $^{131}_{53}I$ emits β and γ radiation.

State two differences between β and γ radiation. [2]

A β, fast-moving electrons; γ, electromagnetic waves

β, (negative) charge; γ, uncharged

Q During its stay in the clinic the cat cannot be given much 'love and care'. Explain this statement. [1]

A The cat emits ionising radiation, which is hazardous to people working at the clinic.

In questions such as this, make sure you concentrate on the physics aspects. For example you would not get a mark just for explaining that 'love and care' might mean stroking the cat!

Q A safe level of radiation is 50 MBq. Calculate how long the cat must stay in the clinic before it is allowed home. [4]

A

$$t_{1/2} = 8 \text{ days} = 8 \times 24 \times 60 \times 60 \text{ s}$$

$$\lambda = \frac{\ln(2)}{t_{1/2}} = \frac{\ln(2)}{(8 \times 24 \times 60 \times 60 \text{ s})} = 1.0 \times 10^{-6} \text{ s}^{-1}$$

$$A = A_0 e^{-\lambda t}$$

so

$$\ln\left(\frac{A_0}{A}\right) = \lambda t$$

$$t = \frac{\ln(80 \text{ MBq} / 50 \text{ MBq})}{(1.0 \times 10^{-6} \text{ s}^{-1})}$$

$$= 4.7 \times 10^5 \text{ s} \quad (= 5.4 \text{ days})$$

Q State one assumption you made in the calculation above. [1]

A The cat does not lose (excrete) any $^{131}_{53}\mathrm{I}$

or The daughter product of $^{131}_{53}\mathrm{I}$ is not radioactive.

? *Quick check questions*

1 The isotope $^{16}\mathrm{N}$ decays by emission of a beta particle ($^{0}_{-1}\mathrm{e}$) to produce an isotope of oxygen (atomic number 8). Write an equation for this decay.

2 The half-life of $^{16}\mathrm{N}$ is 7.1 s.

a Calculate the decay constant for $^{16}\mathrm{N}$.

b Show that the time taken for the mass of $^{16}\mathrm{N}$ in the sample to decrease from 5.0 μg to 1.0 μg is approximately 16 s.

c What is the activity of the sample when the mass of $^{16}\mathrm{N}$ is 5.0 μg?

Nuclear processes

STA Section 2.3

The mass of any nucleus is slightly less than the total mass of the separate nucleons. This difference is called the **mass defect** Δm.

Imagine the separate nucleons coming together to form a nucleus. As they become bound together they *lose energy*. This *lost energy* is called the **binding energy** ΔE and is related to the mass defect:

$$\Delta E = c^2 \Delta m$$

where c is the speed of light (3.00×10^8 m s^{-1}). To pull a nucleus apart into its separate nucleons, this energy ΔE must be supplied.

Binding energy per nucleon

The binding energy per nucleon of a particular nucleus is calculated by dividing the total binding energy by the number of nucleons. You can think of this as the energy needed to remove one nucleon from the nucleus.

Nuclear masses are often expressed in atomic mass units, u. 1 u = 1.67×10^{-27} kg.

Proton mass m_p = 1.00728 u. Neutron mass m_n = 1.00867 u.

Binding energies are often expressed in electronvolts, eV, or MeV.
1 MeV = 1.60×10^{-13} J.

Worked example

The nucleus $^{10}_{5}$B has mass 10.01294 u. Calculate its mass defect and binding energy. Express the binding energy in MeV per nucleon.

Step 1 $^{10}_{5}$B has proton number 5, so it contains five protons. It has 10 nucleons in total so it must also have five neutrons.

$$\textbf{mass of separate nucleons} = 5\,m_p + 5\,m_n$$

$$\textbf{mass defect } \Delta m = 5 \times 1.00728 \text{ u} + 5 \times 1.00867 \text{ u} - 10.01294 \text{ u} = 0.06681 \text{ u}$$

$$= 0.06681 \times 1.67 \times 10^{-27} \text{ kg} = 1.12 \times 10^{-28} \text{ kg}$$

Keep all the significant figures until after you have done the subtraction.

Step 2 Calculate the binding energy

$$\Delta E = c^2 \Delta m = (3.00 \times 10^8 \text{ m s}^{-1})^2 \times 1.12 \times 10^{-28} \text{ kg}$$

$$= 1.00 \times 10^{-11} \text{ J}$$

Step 3 Convert this to electronvolts:

$$\Delta E = \frac{1.00 \times 10^{-11} \text{ J}}{1.60 \times 10^{-13} \text{ J MeV}^{-1}} = 62.8 \text{ MeV}$$

Binding energy per nucleon of $^{10}_{5}$B = 62.8 MeV/10 = 6.28 MeV

The figure shows how the binding energy per nucleon depends on nucleon number. The nucleus of $^{56}_{26}F$ requires the most energy to remove a nucleon. Its nucleons are most tightly bound together.

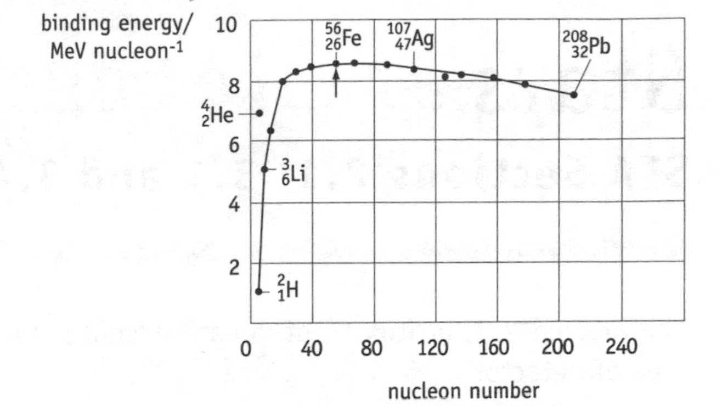

Fusion

When two light nuclei undergo **fusion** they join to make a single nucleus. The average binding energy per nucleon increases and the nucleus loses mass.

As nuclei have positive charge, a lot of energy is needed in the first place to push them together against the force of electrostatic repulsion. For nuclear fusion to take place, temperatures of about 10^7 K or higher are needed to provide the nuclei with enough kinetic energy, and high densities are needed so that they collide frequently. These conditions are found in the cores of stars.

Fission

Nuclear **fission** involves a heavy nucleus splitting to form two lighter nuclei (and usually some neutrons). The average binding energy per nucleon increases so there is an overall loss of mass.

In both fission and fusion, the change in binding energy leads to an increase in kinetic energy and/or the emission of a photon.

✓ *Quick check 1–3*

? *Quick check questions*

1 The nucleus $^{54}_{26}Fe$ has mass 53.93962 u. Calculate its binding energy in MeV per nucleon.

2 For each of the following statements, say whether it applies to **A** nuclear fusion, **B** nuclear fission, **C** both, or **D** neither.

- **a** Two light nuclei join together.
- **b** One heavy nucleus splits apart.
- **c** Leads to an overall loss of mass.
- **d** Leads to an increase in binding energy per nucleon.
- **e** Leads to an overall reduction in binding energy.
- **f** Leads to an increase in kinetic energy and/or emission of radiation.
- **g** Requires high temperature and density.

3 The main nuclear fusion reactions at the Sun's core are summarized by this equation:

$$4\,{}^{1}H \rightarrow {}_{2}He + 2\,{}^{0}e + 2\bar{\nu}_e$$

- **a** Fill in the missing numbers.
- **b** Calculate the energy released by the fusion of 1 kg of hydrogen nuclei (6.02×10^{26} nuclei). (Mass of He nucleus is 4.00260 u. Treat the e and ν as having zero mass.)

Stars

STA Sections 2.1, 3.1 and 3.4

The **intensity**, I, or **flux**, F, of radiation from a star is the power received per unit area of detector.

If the distance, d, to a star is known then its **luminosity** can be calculated from its flux using an inverse-square law:

$$F = \frac{L}{4\pi d^2}$$

A star's luminosity, L, is the total power it emits. It is usually expressed as a multiple of the Sun's luminosity, L_{Sun}.

Worked example

At the top of the Earth's atmosphere, the intensity of the Sun's radiation is about 1.4 kW m^{-2}. The distance from Earth to Sun is 1.50×10^{11} m. Show that the Sun's luminosity is about 4×10^{26} W.

Use the inverse-square law:

$$L = 4\pi d^2 F = 4\pi \times (1.50 \times 10^{11}\ \text{m})^2 \times 1.4 \times 10^3\ \text{W m}^{-2}$$

$$= 3.96 \times 10^{26}\ \text{W}$$

✓ Quick check 1

Properties of stars

The colour of light from a star can be used to estimate its surface temperature. Cool stars appear red and hot stars appear blue-white.

A **Hertzsprung–Russell diagram** (HR diagram) is used to display the properties of stars.

Note that both axes on an HR diagram have logarithmic scales, and that temperature increases to the left. The star marked A is very hot and very luminous.

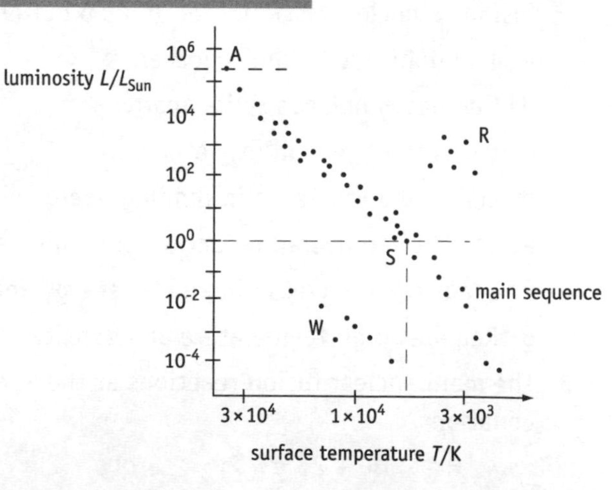

Be careful reading the log scale. Halfway along between powers of ten does *not* correspond to 5×10^x.

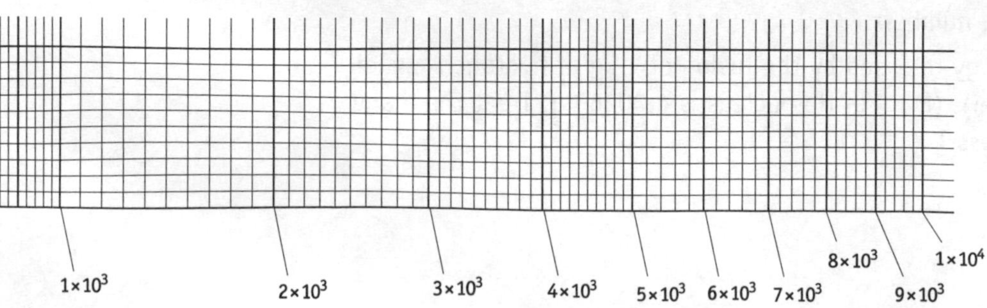

✓ Quick check 2

- Most stars appear within the diagonal band known as the **main sequence**. In these stars, energy is released by the fusion of hydrogen into helium. The Sun, S, is a main-sequence star.
- The stars marked R are **red giants**. They are quite cool but very large so they are very luminous. These stars are powered by the fusion of helium and heavier nuclei.
- The stars labelled W are **white dwarfs**. They are very hot, but their small size means that they emit very little light. They have reached the end of their lives and there is no fusion taking place within them.

Stars form from huge clouds of gas and dust. They join the main sequence soon after they are formed. They settle to an almost-constant luminosity and temperature while hydrogen fusion takes place. Eventually the hydrogen in the core is all turned to helium, the star swells, cools and becomes a red giant. After fusion of heavier nuclei stops, the star cools and shrinks to become a white dwarf.

Quick check questions

1. A white dwarf has luminosity $L = 10^{-3} L_{Sun}$. At the top of the Earth's atmosphere, the Sun's flux is 1.4 kW m^{-2} and the white dwarf's flux is just 7.0×10^{-13} W m^{-2}. Calculate the distance to the white dwarf as a multiple of the Sun's distance.
2. From the figure on page 60, estimate the luminosity of the star labelled A. Give your answer as a multiple of the Sun's luminosity.

Gravitation

STA Section 3.2

Two masses m and M whose centres separated by a distance r attract each other with a force, F, that is described by an inverse-square law:

$$F = \frac{GMm}{r^2}$$

where G is the universal gravitational constant. $G = 6.67 \times 10^{-11}$ N m^2 kg^{-2}.

Worked example

Calculate the gravitational attraction between Saturn and the Sun.
(Sun's mass $M = 1.99 \times 10^{30}$ kg. Saturn's mass $m = 5.69 \times 10^{26}$ kg. Radius of Saturn's orbit $r = 1.43 \times 10^{9}$ km.)

Use the equation $F = GMm/r^2$:

$$r = 1.43 \times 10^9 \text{ km} = 1.43 \times 10^{12} \text{ m}$$

$$F = \frac{6.67 \times 10^{-11} \text{ N m}^2 \text{ kg}^2 \times 1.99 \times 10^{30} \text{ kg} \times 5.69 \times 10^{26} \text{ kg}}{(1.43 \times 10^{12} \text{ m})^2}$$

$$= 3.69 \times 10^{22} \text{ N}$$

Circular orbits

If an object mass m is in circular orbit around an object mass M, then the gravitational force provides the necessary centripetal acceleration:

$$a = \frac{F}{m} = \frac{v^2}{r}$$

The orbital speed, v, is related to the orbital period, T:

$$v = \frac{2\pi r}{T}$$

These relationships can be used to relate the period of an orbit to its radius, e.g. for planets orbiting the Sun, or satellites orbiting a planet.

$$T^2 = \frac{4\pi^2 r^3}{GM}$$

✓ Quick check 1

Gravitational field strength

The strength, g, of a **gravitational field** is defined as the force acting on unit mass placed at a point in the field. Around a point mass M, or a spherical mass, g follows an inverse-square law:

$$g = \frac{F}{m} = \frac{GM}{r^2}$$

At the Earth's surface, $g = 9.81\ \mathrm{N\ kg^{-1}}$.

Gravitational fields always produce attractive forces. They are detected by mass. Electric fields, on the other hand, can produce either attractive or repulsive forces. They are detected by charge. The electric field around a point or a sphere of charge follows an inverse-square law.

✓ Quick check 2–4

? Quick check questions

1 The Moon's orbital period is 27.3 days and its orbital radius is 3.8×10^5 km. Calculate the mass of the Earth.

2 For each of the following statements, say whether it applies to **A** gravitational fields, **B** electric fields, or **C** both.

 a Give rise to a force that depends only on an object's charge.

 b Give rise to a force that depends only on an object's mass.

 c Around a spherical object the field strength obeys an inverse-square law.

 d The field strength is close to zero at a very large distance from any object.

 e Always produces an attractive force.

 f Can produce an attractive or a repulsive force.

3 Using the Earth's mass (question **1**) and surface gravitational field strength, calculate:

 a the Earth's radius

 b the average density of the Earth.

4 The gravitational field strength at the Earth's surface is g. Which of the following is the gravitational field strength at the surface of a planet which is half as dense as Earth and whose diameter is four times that of the Earth?

A $2g$, **B** $4g$, **C** g, **D** $g/2$, **E** $8g$.

Kinetic theory and gas laws

STA Section 3.3

Molecules have kinetic energy because they are moving. They also have potential energy because there are attractive forces between them. Increasing the separation of molecules increases their potential energy. The sum of molecular kinetic and potential energies within a sample of matter is known as its **internal energy.**

In any collection of molecules, kinetic and potential energy are shared at random between all the molecules. There is a range of speeds and hence a range of kinetic energies. The average molecular kinetic energy is proportional to absolute temperature, *T*. In a gas,

$$\tfrac{1}{2} m \langle c^2 \rangle = \frac{3kT}{2}$$

where *k* is the Boltzmann constant. $k = 1.38 \times$ J K^{-1}. $\langle c^2 \rangle$ is the **mean square speed,** which is the result of squaring all the molecular speeds, then dividing by the total number of molecules.

Absolute temperature is measured in kelvin, K.

temperature in K = temperature in °C + 273

A temperature *difference* of 1 K is exactly the same as a *difference* of 1 °C.

When $T = 0$ K = −273 °C, molecular kinetic energy becomes zero. This temperature is known as **absolute zero.**

Change of state

When a substance melts or evaporates, an energy input is required to increase the molecule potential energy but there is no increase in kinetic energy and hence no rise in temperature. When the substance condenses or solidifies, energy is released as the molecular potential energy decreases but the temperature does not fall.

✓ Quick check 1, 2

Worked example

What is the mean square speed of oxygen molecules at room temperature $T = 300$ K? What is the square root of this value (known as the root-mean-square or r.m.s. speed)? The mass of an oxygen molecule is $32 \times 1.67 \times 10^{-27}$ kg.

Step 1 Use the equation for average kinetic energy:

$$\tfrac{1}{2} m \langle c^2 \rangle = \frac{3kT}{2} \quad \text{so} \quad \langle c^2 \rangle = \frac{3kT}{m}$$

Step 2 Substitute and solve:

$$\langle c^2 \rangle = \frac{3 \times 1.38 \times 10^{-23}\ \text{J K}^{-1} \times 300\ \text{K}}{32 \times 1.67 \times 10^{-27}\ \text{kg}}$$
$$= 2.32 \times 10^5\ \text{m}^2\ \text{s}^{-2}$$

Step 3 Take the square root to find the r.m.s. speed:

$$\text{r.m.s. speed} = \sqrt{\langle c^2 \rangle} = 482\ \text{m s}^{-1}$$

An ideal gas

The relationship between pressure, p, volume, V, and absolute temperature, T, of a gas can be approximately described by the **ideal gas equation:**

$$pV = nRT$$

where n is the number of moles present and R the **molar gas constant.** One mole of molecules (or atoms) contains 6.02×10^{23} molecules (or atoms). $R = 6.02 \times 10^{23} \times k = 8.31\ \text{J K}^{-1}\ \text{mol}^{-1}$.

✓ Quick check 3

An **ideal gas** is one that obeys the equation exactly. In practice a gas with low density and high temperature behaves very like an ideal gas.

For a fixed amount of gas (which means a fixed number of moles)

$$\frac{p_2V_2}{T_2} = \frac{p_1V_1}{T_1}$$

where the subscripts denote p, V and T before and after a change has occurred.

Worked example

A closed container of gas is initially at atmospheric pressure (1.00×10^5 Pa) and a temperature of 300 K. What is the pressure in the container when it is heated to 420 K?

Step 1 The volume is fixed so $V_2 = V_1$. Hence

$$\frac{p_2}{T_2} = \frac{p_1}{T_1} \quad \text{and} \quad p_2 = p_1T_2/T_1$$

Step 2 Substitute and solve:

$$p_2 = 1.00 \times 10^5\ \text{Pa} \times \left(\frac{420\ \text{K}}{300\ \text{K}}\right) = 1.40 \times 10^5\ \text{Pa}$$

✓ Quick check 4

Worked exam question

A bottle half full of cool, sparkling apple juice has a special stopper which rests on the neck of the bottle and helps to prevent the juice losing its 'fizz'. From time to time, the stopper lifts slightly, releasing a small amount of gas, then falls back.

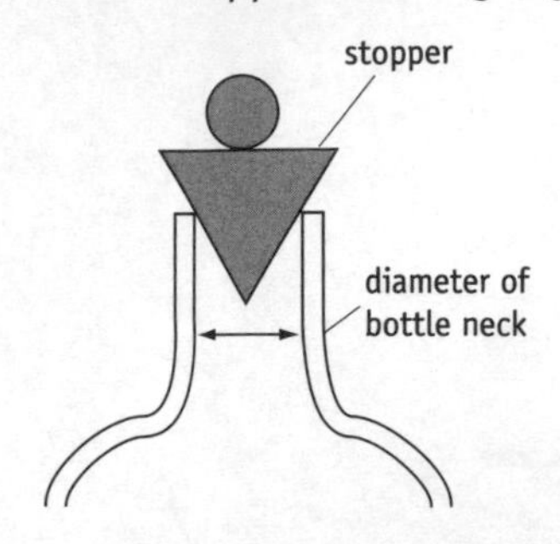

The diagram shows the stopper resting on the neck of the bottle.

Q Draw labelled arrows on the diagram to represent the vertical components of the forces acting on the stopper as it lifts. [2]

A The arrows must be labelled 'weight' and 'upward force (due to gas pressure).

The examiners did not accept just 'gravity' as the downward force or just 'pressure' as the upward force. Pressure is related to force but it is not the same thing.

Q Show that the excess pressure of the gas in the bottle just before the stopper lifts is about 5 kPa.

(Mass of stopper = 0.12 kg; diameter of bottle neck = 1.8×10^{-2} m.) [3]

A Excess pressure p = force to lift stopper/area = weight of stopper/area

area = πr^2 and radius r = diameter/2 = 9×10^{-3} m

So

$$\mathbf{p = \frac{0.12\ kg \times 9.81\ N\ kg^{-1}}{\pi \times (9 \times 10^{-3}\ m)^2}}$$

$$\mathbf{= 4.6 \times 10^3\ Pa}$$

Q By estimating the volume and temperature of the bottle, calculate an approximate value for the quantity of gas, in moles, in the bottle just before the stopper lifts. (Atmospheric pressure = 100 kPa.) [5]

A $\mathbf{pV = nRT}$ so $\mathbf{n = pV/RT}$

You can expect to find a question on the PSA5 paper which asks you to estimate quantities. When you are asked for an estimate, any reasonable values are awarded marks.

For volume you might choose a value anywhere between about 0.1 litre (a small cupful) and 2 litres (a large lemonade bottle).

V = 0.2 litre

1 m^3 = 10^3 litre so $V = 0.2 \times 10^{-3}$ m^3

The juice is 'cool' so its temperature will be less than a typical room temperature (25 °C) but above freezing (0 °C).

T = 10 °C = 283 K

Volume must be in m^3. Take care with the conversion. 1 litre = 0.1 m × 0.1 m × 0.1 m = 1×10^{-3} m^3.

Remember to express T in kelvin.

Pressure inside bottle, p = atmospheric pressure + excess pressure

$$p = 1.00 \times 10^5 \text{ Pa} + 4.6 \times 10^3 \text{ Pa} = 1.046 \times 10^5 \text{ Pa}$$

Hence

$$n = \frac{1.046 \times 10^5 \text{ Pa} \times 0.2 \times 10^{-3} \text{ m}^3}{8.31 \text{ J mol}^{-1} \text{ K}^{-1} \times 283 \text{ K}}$$

$$= 9 \times 10^{-3} \text{ moles}$$

- *Your answer will depend on the estimates you have made. Provided you have carried out the calculation correctly, you will score marks regardless of whether your estimates were reasonable.*
- *As the quantities are only rough estimates, you cannot justify more than one significant figure in the final answer. You would lose a mark for writing more than two significant figures.*

? Quick check questions

1. Describe the changes in internal energy when water is heated from room temperature then boils to steam.
2. An ideal gas is heated from 300 K to 600 K. By what factor does the average speed of its molecules change?
3. How many moles are there in 2.24×10^7 cm^3 of an ideal gas at 295 K and 1.0×10^5 Pa?
4. Explain why it is often difficult to reopen a freezer door immediately after it has been closed.

The universe

STA Part 4

Light from stars and galaxies contains *emission and absorption lines* – certain wavelengths where the light is much brighter or fainter than at other wavelengths. The lines depend on which atoms, molecules and ions are present. Nearby galaxies all have a very similar pattern of lines.

Redshift

Light from distant galaxies contains lines whose wavelengths are stretched compared with those of nearby galaxies. The stretch, or **redshift**, z, is related to a galaxy's speed, v. If v is much less than c, the speed of light:

$$z = \frac{\Delta\lambda}{\lambda} = \frac{v}{c}$$

where λ is the original wavelength and $\Delta\lambda$ is the difference between the original wavelength and the wavelength observed.

Measurements of distance and redshift show that the greater the distance, the greater the redshift. Galaxies are moving apart, and the universe is expanding uniformly.

Hubble's law

The relationship between recession speed and distance, d, is known as **Hubble's law**. The gradient of the graph to the right gives the **Hubble constant**, H_0.

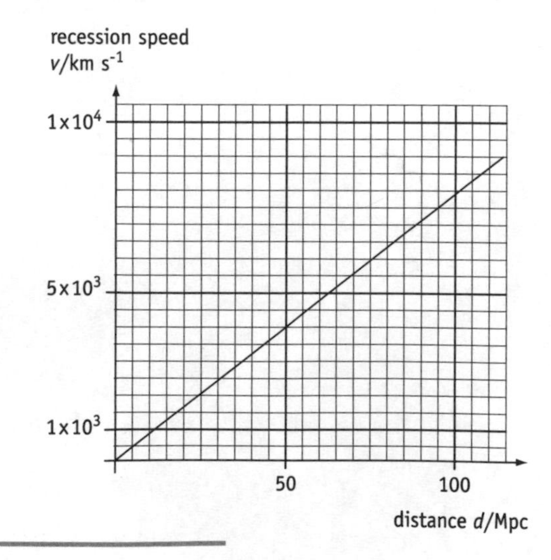

$$v = H_0 d$$

Astronomical distances are commonly expressed in parsecs (pc) or Mpc and recession speeds in km s^{-1}, so H_0 has units km s^{-1} Mpc^{-1}.

$$1\text{pc} = 3.09 \times 10^{16}\ \text{m}.$$

Worked example

Light from a distant galaxy has an emission line with wavelength 445 nm. The pattern of lines indicates that the line has wavelength 410 nm 'in the lab'. Using H_0 = 75 km s^{-1} Mpc^{-1}, calculate the galaxy's redshift, recession speed and distance.

Step 1 Calculate the redshift, z:

$$\Delta\lambda = 445\ \text{nm} - 410\ \text{nm} = 35\ \text{nm}$$

$$z = \frac{\Delta\lambda}{\lambda} = \frac{35\,\text{nm}}{410\,\text{nm}} = 0.0854$$

Step 2 Use this value to find v:

$$v = cz = 0.085 \times 3.00 \times 10^8 \text{ m s}^{-1}$$

$$= 2.56 \times 10^7 \text{ m s}^{-1} = 2.56 \times 10^4 \text{ km s}^{-1}$$

Step 3 Use Hubble's law to find d:

$$d = \frac{v}{H_0} = \frac{2.56 \times 10^4 \text{ km s}^{-1}}{75 \text{ km s}^{-1} \text{ Mpc}^{-1}} = 341 \text{ Mpc}$$

Take care with units. Put v into km s^{-1}, then with H_0 in km s^{-1} Mpc^{-1} you get d in Mpc.

✓ Quick check 1

Expansion of the universe

From Hubble's law we can deduce the time taken for galaxies to get where they are now. This is the time since the beginning of the expansion – the beginning of the universe, known as the Big Bang. If we assume that the universe has always been uniformly expanding at its current rate, then

$$\text{age of universe} = \frac{d}{v} = \frac{1}{H_0}$$

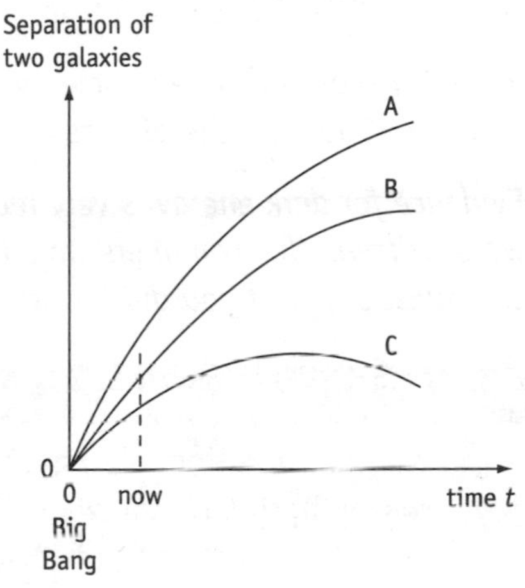

In fact gravitational forces will have slowed the expansion. There are three possible futures, which depend on the amount of mass in the universe and therefore the magnitude of the gravitational forces. These are marked on the figure.

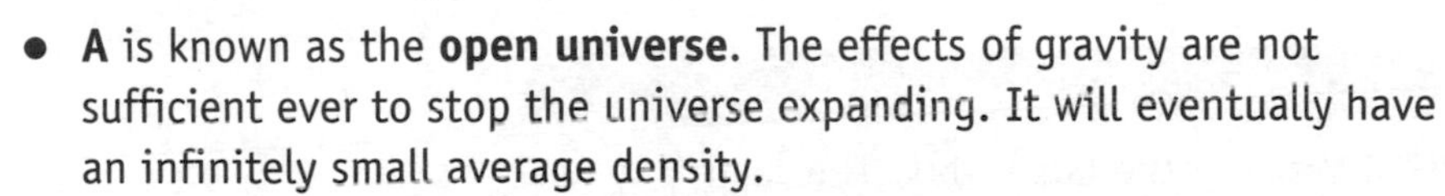

- **A** is known as the **open universe.** The effects of gravity are not sufficient ever to stop the universe expanding. It will eventually have an infinitely small average density.

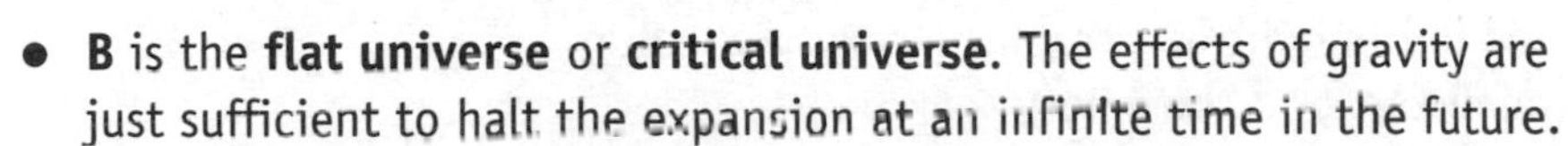

- **B** is the **flat universe** or **critical universe.** The effects of gravity are just sufficient to halt the expansion at an infinite time in the future.

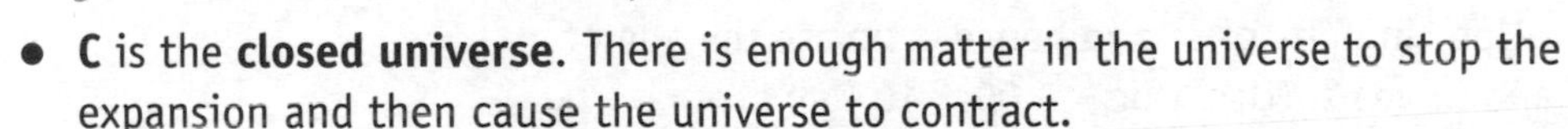

- **C** is the **closed universe.** There is enough matter in the universe to stop the expansion and then cause the universe to contract.

At present, estimates of the amount of ordinary matter in the universe suggest an open universe. However, there is evidence for matter that cannot be detected using the emission or absorption of radiation. This **dark matter** might be enough to produce a closed universe.

✓ Quick check 2, 3

Worked exam question

Q Discuss the ultimate fate of the universe. Your answer should include reference to dark matter and reasons why the fate of the universe is uncertain. [6]

- *You can expect to find an 'open-ended' question such as this on the PSA5 paper.*
- *Pause and plan your answer before you start writing. Be guided by the number of marks. Here there are six marks so you should aim to say at least six things that are relevant.*
- *Your answer can be quite sort and still gain full marks. A long rambling answer will not necessarily score high marks. A good answer should be able to fit into the space provided.*

- *The examiners' mark scheme usually contains more points than are necessary for full marks, so different people might write different answers and still gain full marks. The points listed here are those that the examiners might expect to see. Other correct relevant points would also gain marks.*

A
- The universe may continue to expand
- or may collapse back on itself.
- The future depends on the amount of matter in the universe
- because mass leads to gravitational forces/deceleration of galaxies.
- So far, not enough ordinary matter has been found to stop the expansion.
- Dark matter is matter that cannot be detected using radiation
- but it still exerts a gravitational force.
- Examples of dark matter might include:
  - neutrinos
  - WIMPS (weakly interacting massive particles).

Recently there has been some evidence for 'dark energy'that accelerates the expansion despite the effects of gravity.

Evidence for dark energy is very recent and has arisen since the SHAP course books were written. The examiners would not expect you to know about it but it is relevant to the question, so if you did include it you would gain marks.

? Quick check questions

1 An emission line has wavelength 434 nm in the laboratory. The same line has wavelength 512 nm when observed in the light of a distant galaxy. Assuming $H_0 = 75$ km s^{-1} Mpc^{-1}, calculate the distance to the galaxy.

2 a Express $H_0 = 75$ km s^{-1} Mpc^{-1} in units of s^{-1} and hence estimate the age of the universe in seconds and in years. (Use 1 pc = 3.09×10^{16} m; 1 year = 3.16×10^{7} s.)

b Explain why this is an *estimate*.

3 a Write a short paragraph explaining the meaning of the phrase 'closed universe'.

b Explain how astronomical observations in the very distant future in a closed universe would differ from those of today.

Practice exam questions for PSA5

1 Good suspension in a car helps prevent resonance in the various parts of the car such as the seats and mirrors. Each part has its own frequency of vibration.

What is this frequency called? [1]

Explain the term resonance. [2]

Different frequencies of vibration can be applied to the system. Sketch a graph to show the variation of amplitude with forcing frequency for a vibrating system. Label your graph line A.

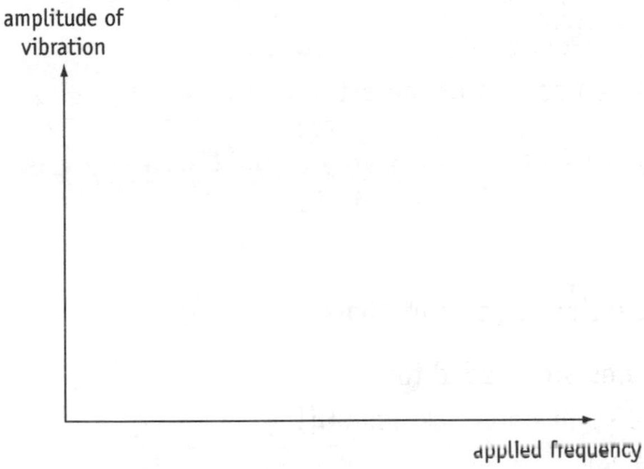

Mark the resonant frequency on the forcing frequency axis with a label R.

Add a second line to show the effect of additional damping on this system. Label this line B. [4]

How does good suspension in a car help prevent resonance in the various parts of the car? [2]

From Edexcel June 2001, Unit PSA6i [Total: 9]

2 Active noise control systems are based on the physics of destructive interference. Background noise is sampled and fed back into the system in such a way that the original noise is totally removed or at least significantly reduced. A typical system is shown in the diagram below.

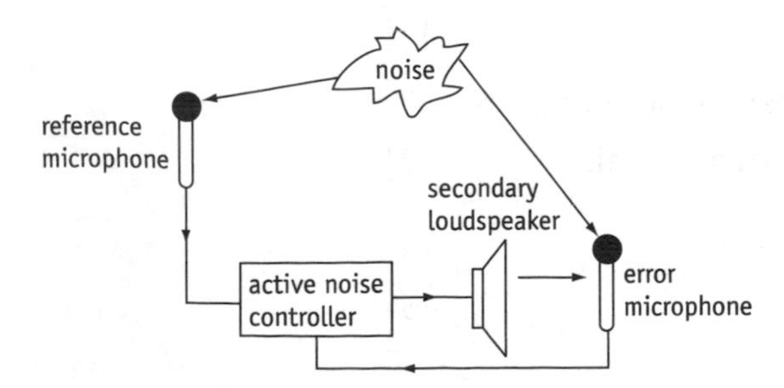

In the boxes below add sketches of the waveforms that could be expected from the secondary loudspeaker and at the error microphone when the sound is **a** totally removed, and **b** significantly reduced. [4].

	sound totally removed	sound significantly reduced
reference microphone		
secondary loudspeaker		
error microphone		

Explain why the signal from the error microphone is required as part of the active noise control system. [2]

From Edexcel June 2000, Unit PSA6i [Total: 6]

3 Scientists in the UK have achieved a controlled fusion reaction for a short time.

Explain why high densities of matter and high temperatures are needed for nuclear fusion. [2]

Why is it difficult to achieve

(i) high density

(ii) high temperature? [2]

In one fusion reaction two deuterium ($^{2}_{1}H$) nuclei combine to form a helium nucleus ($^{3}_{2}He$). Write an equation for this reaction. [1]

The masses involved are:

	mass/u
^{2}H	2.01410
^{3}He	3.01603
neutron	1.00867

$1 \text{ u} = 1.67 \times 10^{-27}$ kg

Calculate the energy released in this reaction. [4]

Hence calculate the energy released when 1.0 kg of deuterium nuclei fuse to form $^{3}_{2}He$. (1.0 kg of deuterium contains 3.0×10^{26} deuterium nuclei.) [2]

Describe one environmental advantage of fusion over fission. [1]

From Edexcel June 2000, Unit PSA6i [Total: 12]

4 The mission for NEAR (Near Earth Asteroid Rendezvous spacecraft) is to go into orbit 35 km from the centre of the asteroid Eros. Estimate the required orbital speed. [8]

Eros is a potato-shaped asteroid, 33 km long and 13 km diameter.

Each 1 m^3 of Eros rock has a mass of approximately 2700 kg.

(You could start by estimating the mass of Eros and then go on to derive an expression for orbital speed using expressions for gravitational attraction and centripetal force.)

Give one reason why it may be difficult to achieve the correct orbit after a 3-year space flight from Earth to Eros at a speed of about 30 km s^{-1}. [1]

From Edexcel June 2000, Unit PSA6i [Total: 9]

5 Some meteorites found on Earth are thought to have originated from the Moon – knocked off the lunar surface by the impact of a larger object. These meteorites can be dated using the radioactive decay of potassium-40 to stable argon-40. The argon is trapped in the meteorite. The masses of potassium-40 and argon-40 can be measured in a sample of the meteorite.

The half-life of potassium-40 is 1.3×10^9 years.

Explain what is meant by the half-life of a radioactive isotope. [1]

A sample contains 3.0×10^{15} atoms of potassium-40. Calculate the activity, in Bq, of the sample. [3]

When measuring this activity in a laboratory on Earth what difficulty would be encountered?

How could this difficulty be overcome? [3]

The sample also contains 2.1×10^{16} atoms of argon-40. Calculate the age of the meteorite. State any assumptions that you made. [4]

From Edexcel June 2000, Unit PSA6i [Total: 11]

6 In 1960 a brilliant physicist called Richard Feynman gave an interesting lecture entitled 'there's always room at the bottom'. He discussed the idea that we should eventually be able to stick individual atoms together to make useful new materials and build objects on a very small scale. This has become known as *nanotechnology* and the paragraph below illustrates a possible application that may become a reality in the next 20 or 30 years.

If people suffer from very poor circulation, tissue can become damaged. A temporary solution would be to replicate one of the functions of red blood cells by manufacturing tiny spheres full of compressed oxygen and inject these directly into the blood stream. These spheres could then slowly release their oxygen.

Nanotechnology offers the promise of making extremely strong, diamond-like materials in any shape required. The spheres would be mass-produced very cheaply. Spheres of internal diameter 1.0×10^{-7} m could be filled with oxygen to a pressure of 1.0×10^{8} Pa.

State the meaning of the word *nano* when used as a prefix in front of a unit. [1]

The spheres are filled with oxygen at a body temperature of 310 K. Calculate the number of moles of oxygen in one sphere. Assume oxygen behaves as an ideal gas. [4]

The typical oxygen consumption of an adult is 2.5×10^{-6} m^3 per minute at atmospheric pressure (1.0×10^{5} Pa). Calculate the volume in cm^3 of spheres required to sustain the oxygen requirement of an adult for one hour. Assume the volume of material used for the sphere is negligible compared with its internal volume. [3]

From Edexcel June 2000, Unit PSA6ii [Total: 8]

7 The diagram below shows the spectrum of light from the quasar 3C273 (a quasar is a very luminous type of galaxy). The four peaks marked are hydrogen lines that have shifted in wavelength. In the laboratory these lines have wavelengths 410, 434, 486 and 656 nm.

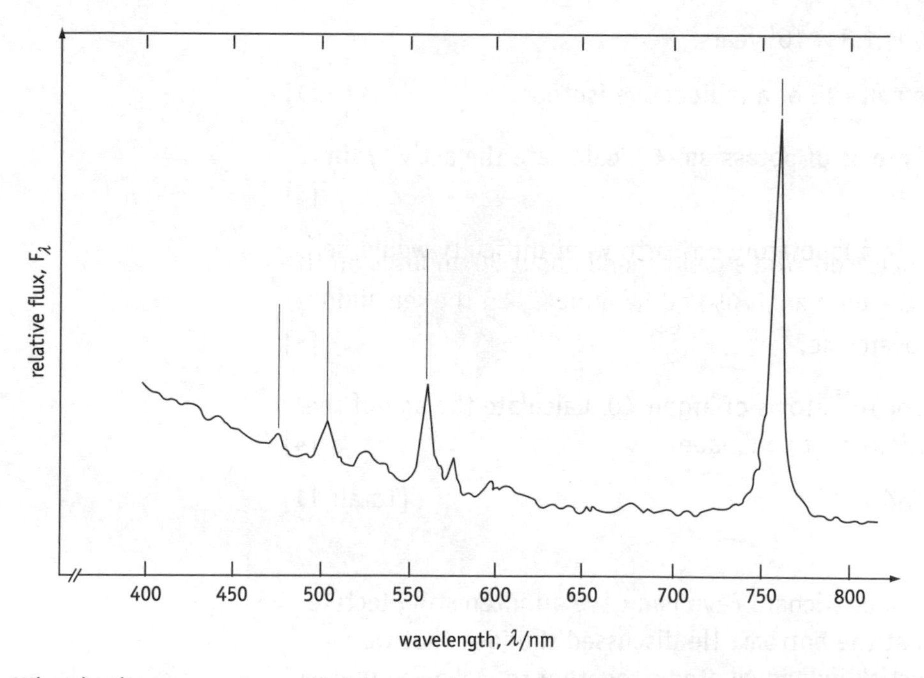

What is the name given to this shift in wavelength? [1]

Without doing a calculation, what does this spectrum allow you to deduce about the motion of 3C273? [1]

Using any one line on the diagram, calculate the speed of 3C273 relative to Earth. [3]

Assuming $H_0 = 75$ km s^{-1} Mpc^{-1}, calculate the distance of 3C273. [2]

What observations would you expect to make of light from a galaxy twice as distant as 3C273? [2]

From Edexcel Specimen paper, Unit PSA6i [Total: 9]

Unit PSA6: Exploring Physics

The synoptic paper

The synoptic paper ranges over the entire course and you can expect to find questions on any physics topic from any of the AS or A2 units. Within any one question, you can expect to find more than one area of physics, sometimes from quite different parts of the course. In the worked examples and questions used in this guide you will find references to the relevant section(s) of your physics textbooks.

The best way to prepare for the synoptic paper is to revise all of your AS and A2 work thoroughly, so any revision that you do for the other papers will also be very helpful for the synoptic test. Even more than the other papers, the synoptic paper tests your understanding of physics rather than your ability to memorise and recall. You need to be able to spot the relevant physics in some unfamiliar situations and to apply your knowledge and understanding to some often quite demanding problems.

As well as revising your physics, you need to become familiar with the format of the synoptic test paper, which is a bit different from the other SHAP papers. The first part is always a comprehension exercise, and the second is a data analysis. Then there are two other questions which are in a similar style to those on the PSA4 and PSA5 papers; at least one of these is a structured question and the other is usually a more open-ended unstructured question. You should spend about 30 minutes on the comprehension, 30 minutes on the data analysis and 30 minutes on the remaining questions.

Comprehension

This is based on a short scientific article, usually about one page long. The passage serves as a stimulus to prompt questions on a range of topics. The total marks for the comprehension are about one-third of the total marks on the paper and you should spend about 30 minutes on this part.

Read the passage through once to get an idea of the topics covered. Don't worry about trying to understand everything in the passage straightaway – you can come back to particular points if you need to when answering the questions. You often need numerical data and other information from the passage so it can be helpful to highlight useful information with a marker pen.

Passage and sample questions

Over a century ago, at Cambridge University, J J Thomson discovered the electron. A flow of electrons is what produces an electric current. The principle behind J J's experiment was to measure the deflection of an electron beam (cathode ray) of known energy in a uniform magnetic field. Now physicists in the same laboratory have finally managed to isolate a single electron, play with it and measure its tiny charge with unprecedented accuracy. This means that physicists have produced a much more reliable way of measuring electric current, a breakthrough that has major implications for the future of electronics.

Scientists have tried to isolate a single electron using smaller and smaller capacitors, which store electric charge. But while they have been able to witness the effect of adding one electron to a group – moving, say, from 100 to 101 electrons – they have not been able to get an electron on its own. Professor Michael Pepper and his group at the Cavendish Physics Laboratory tried a different approach. First they created a two-dimensional sheet of electrons, which is easy to do using a semiconductor chip of gallium arsenide. Then they reduced them to a stream of single electrons using a split-gate technique developed in the 1980s. To isolate a single electron they used a method which involved sound waves.

A progressive sound wave will pull electrons with it; they get stuck in the troughs of the wave. A high-frequency sound wave moving along the electron stream can scoop up one electron and carry it to a measuring instrument. 'This is the nearest thing to being able to pick up an electron in your hand', Professor Pepper said. From the frequency of the sound wave, the scientists know the frequency of the emerging electrons and can measure the current using standard instruments. As the frequency of the sound wave is high, the current produced is quite measurable. The physicists can then deduce the charge of a single electron with great precision.

Q An electron is a fundamental particle. Briefly explain the meaning of the word 'fundamental' in this context and give one other example of such a particle. [2]

A Fundamental particles cannot be broken down into other constituents. ✓

Other examples: neutrino, quark, muon, tau. ✓

▶▶ *PRO Section 2.2*

Q Draw a diagram to show the path of electrons moving in a uniform magnetic field. [2]

A

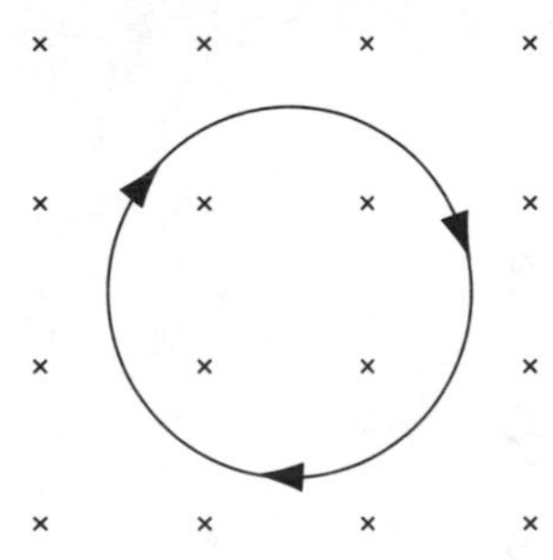

There is a force at right-angles to the field and the direction of motion, which deflects the electrons into a circle. ✓✓

▶▶ *MDM Section 5.3, PRO Sections 4.3 and 4.4*

Q Consider a 2.0 pF capacitor, initially uncharged. Calculate the potential difference across it when 100 electrons are transferred from one plate to the other. [3]

A

$$Q = CV \quad \text{so} \quad V = \frac{Q}{C}$$ ✓

$$V = \frac{100 \times 1.60 \times 10^{-19}\ \text{C}}{2.0 \times 10^{-12}\ \text{F}}$$ ✓

$$= 8.0 \times 10^{-6}\ \text{V}$$ ✓

▶▶ *TRA Section 4.2, MDM Section 4.1*

Q A flat sheet of charge produces a uniform electric field close to each side of the sheet. Sketch the field produced by a flat sheet of electrons. [2]

A

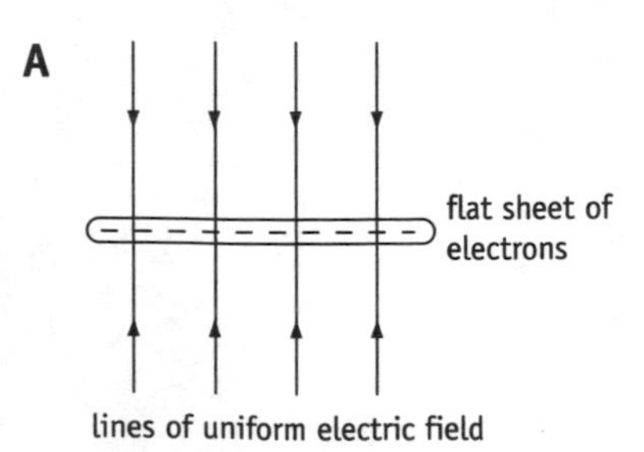

The diagram needs to show:

- equally spaced parallel lines (uniform field) ✓
- arrows on field lines towards sheet (direction of force on positive charge) ✓

▶▶ *MDM Section 5.2*

Q In order to be 'scooped up' by a sound wave (3rd paragraph), electrons need to move at the speed of sound, which is about 340 m s^{-1} in air at room temperature. Calculate the potential difference through which electrons would be accelerated to reach this speed. [3]

A $$eV = \tfrac{1}{2}mv^2 \quad \text{so} \quad V = \frac{\tfrac{1}{2}mv^2}{e}$$ ✓

$$V = \frac{9.11 \times 10^{-31}\ \text{kg} \times (340\ \text{m s}^{-1})^2}{2 \times 1.60 \times 10^{-19}\ \text{C}}$$ ✓

$$= 3.3 \times 10^{-7}\ \text{V}$$ ✓

▶▶ *SPC Section 2.1, MDM Section 5.3*

Q Assuming that just a single electron is 'scooped up' into each wave trough, what would be the frequency of the sound wave needed to produce a current of 1.8 × 10^{-11} A? [4]

A Number of electrons arriving per second = frequency, f ✓

Charge arriving per second = ef, so current $I = ef$. ✓

$$f = \frac{I}{e} = \frac{1.8 \times 10^{-11}\ \text{A}}{1.60 \times 10^{-19}\ \text{C}}$$ ✓

$$= 1.1 \times 10^{8}\ \text{Hz}$$ ✓

 *MUS Section 1.1, SPC Section 2.1*

Q Draw a sketch showing a 'snapshot' of molecules in a gas during the passage of a sound wave and hence suggest how it might be possible for an electron to be 'scooped up' and carried along by such a wave. [3]

A ✓

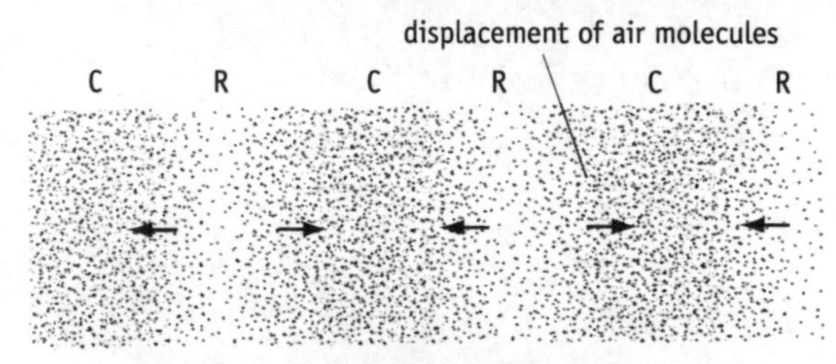

There are alternating regions of high and low pressure. ✓

An electron within a low pressure region might not easily move into a region of higher pressure so could be swept along in the moving 'pressure trough' rather than remaining in its original position. ✓

▶▶ *MUS Section 1.3*

When you are to 'suggest' an explanation, you are given marks for any reasonable explanation based on your knowledge of physics.

Data analysis

In this part you are given some data (often on a topic related to the comprehension passage) and a series of linked questions in which you have to manipulate the data in some way and use a graph. The data analysis carries about one-third of the total marks for the paper, so spend about 30 minutes on this part.

This question tests your ability to draw and interpret graphs and to perform calculations.

When plotting a graph, it is important to pay attention to detail. Marks can be given (or taken off!) depending on the following.

- Choose a scale for each axis that covers at least half the available space and is easy to use – in practice that means using 1, 2, 5 or 10 graph paper squares as the basis for the scale.
- Label each axis with the quantity and its unit.
- Use a *sharp* pencil to plot the points and draw the line.
- Plot each point carefully, using either a *small* dot inside a circle or a *small* cross.
- If the graph is a straight line, draw it using a ruler that is long enough to extend over the whole graph.
- Continue the line beyond the plotted points. Do not force the line to go through any point (such as the origin) just for convenience.

Types of graph

There are four common types of graphical analysis that you can usually expect to find in this part of the paper.

Straight-line graph

The general equation of a straight-line graph is $y = mx + c$. The constant m is equal to the gradient of the graph, and the constant c is the value of y where the line intercepts the y-axis.

If $c = 0$, then the line goes through the origin and y is directly proportional to x. This is not true for any other values of c.

If you are plotting a straight-line graph to test whether two quantities are directly proportional, or to find the value of c, then you must start both axes at $x = 0$ and $y = 0$.

Curved graph

Two common types of curved graph in physics are those showing either exponential decay or an inverse-square relationship. Make sure you know the difference between these!

If decay is exponential, $y = Ae^{-kx}$ and the quantity plotted on the y-axis changes by equal *fractions* for equal *steps* along the x-axis. For example, the number of nuclei of a radioactive isotope halves during each half-life.

In an inverse-square relationship, $y = Ax^{-2}$. Doubling the value of x divides the value of y by 4, multiplying x by 3 divides y by 9, and so on. For example, the intensity or flux of light from a point source falls to one-quarter its original value if you double the distance from the source, and to one-ninth if you multiply the distance by three.

Log–log graph

This type of graph can be used to determine whether two quantities are related by a power law: $y = Ax^f$. If they are, then a graph of log (y) against log (x) obeys the equation

$$\log(y) = \log(A) + p \log(x)$$

The graph is a straight line, with gradient equal to the exponent p and an intercept of log (A) on the vertical axis.

To plot a log–log graph, you can use either base-ten logs (common logs, log or lg on a calculator) or base-e logs (natural logs, ln on a calculator), but it is more usual to use base-ten logs.

Log–linear graph

Use this type of graph to determine whether the relationship between two quantities is exponential: $y = Ae^{kx}$. If it is, then a graph of ln (y) against x obeys the equation

$$\ln(y) = \ln(A) + kx$$

This gives a straight-line graph, with gradient k and intercept ln (A) on the vertical axis.

This type of graph is plotted using base-e logs (natural logs).

Worked examples

1 A number of identical cells are used in a range of different applications. In each case a small *constant* current I gradually reduces the cell's emf to zero. The total discharge time t is different in each application. The current I is measured in each case and the following data recorded.

I/mA	1.0	2.1	3.2	4.0	6.2	7.9	10.1
t/days	25	11	6.7	4.8	2.9	2.0	1.5

Q State the probable sensitivity of the ammeter used to take these measurements. [1]

A To nearest 0.05 mA ✓

For example, the reading listed as 3.2 A must lie between 3.15 and 3.25 mA.

Q It is suggested that I and t are related by a power law of the form $I = kt^n$, where k and n are constants. Draw up a suitable table of values and plot a graph to test this relationship. [7]

A Plot log (I) against log (t). ✓✓

log (I/mA)	0.0	0.32	0.51	0.60	0.79	0.90	1.00
log (t/days)	1.4	1.0	0.83	0.68	0.46	0.30	0.18

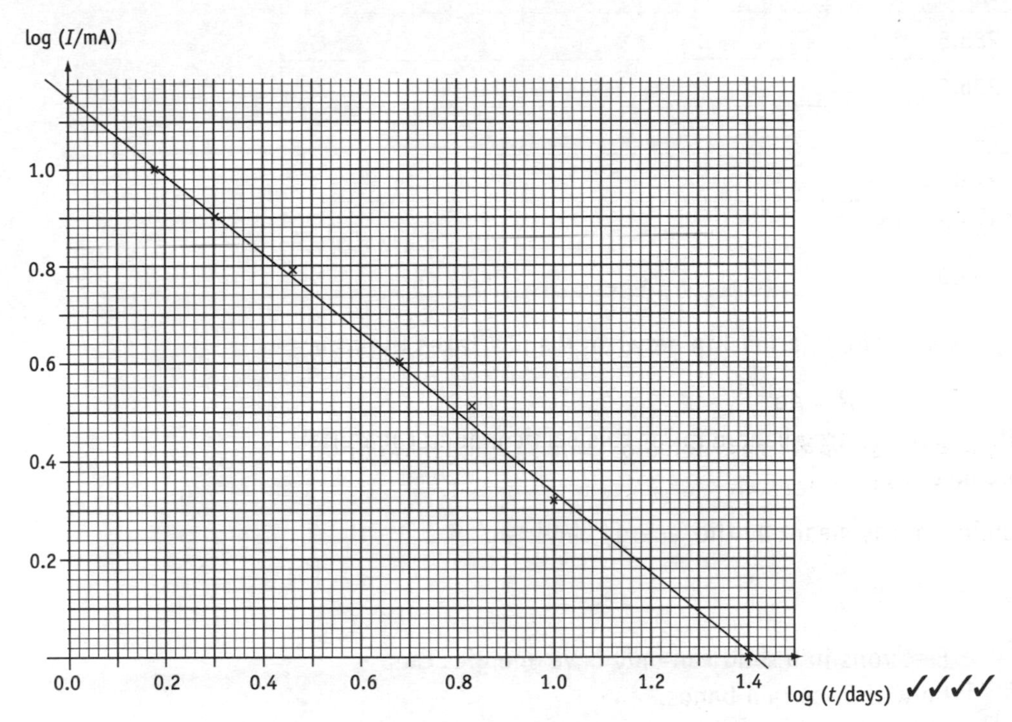

✓✓✓✓

The graph is a straight line so it confirms that the relationship between I and t is a power law as suggested. ✓

Q From your graph, deduce values for n and k. [4]

A n is the gradient of the graph. The line forms its own triangle: 'rise' = −1.15, 'run' = 1.41, so gradient = −1.15/1.41 = −0.82. ✓✓

Note the negative gradient.

The y-axis intercept is log k = 1.15 ✓

k = antilog (1.15) = 14.1 ✓

Q Comment on your value for n. [1]

A The power n is often a 'round number' so here the power is probably −1. ✓

2 The table below shows the resistance, R, of a thermistor at various temperatures, T.

R/Ω	T/K		
4570	283.5		
2398	296.5		
1365	309.5		
631	325.5		
302	345.0		
149	365.0		

Q Simplified theory suggests that the relationship between R and T takes the form:

$$R = Ae^{(E/2kT)}$$

where A is a constant, E is the energy band gap for this semiconductor material and k the Boltzmann constant.

Q Using a diagram, explain what is meant by the 'energy band gap' of a semiconductor. [3]

A

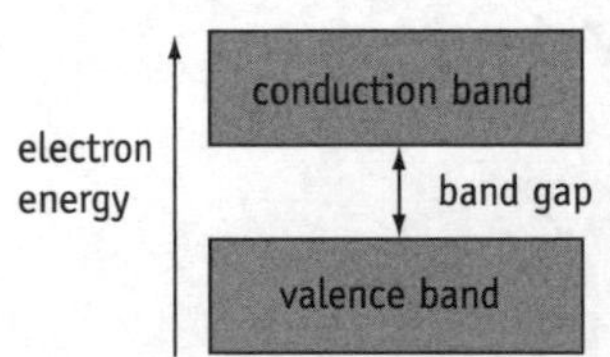

Electrons in a solid can only have energies that lie within certain bands. ✓

Electrons with enough energy to move through the material lie in the conduction band. Electrons with lower energy remain bound within atoms (the valence band). ✓

The gap between these bands is the 'band gap'. ✓

Q Explain briefly why the resistance of a thermistor decreases as temperature rises. [2]

A The higher the temperature, the greater the energy of the electrons. ✓

At high temperatures, more electrons can reach the conduction band, so for a given applied pd there is more current, i.e. resistance decreases. ✓

Q State the precision of the temperature measurement. [1]

A Probably 0.25 K. ✓

For example, the temperature listed as 383.5 K lies between 383.25 K and 383.75 K.

Q Plot a suitable graph to test the suggested relationship. [7]

A Plot a graph of ln (R) against $1/T$. ✓✓✓

R/Ω	T/K	**ln (R/Ω)**	**1/(T/K)**
4570	283.5	8.43	3.53×10^{-3}
2398	296.5	7.78	3.37×10^{-3}
1365	309.5	7.23	3.23×10^{-3}
631	325.5	6.44	3.07×10^{-3}
302	345.0	5.71	2.90×10^{-3}
149	365.0	5.00	2.74×10^{-3}

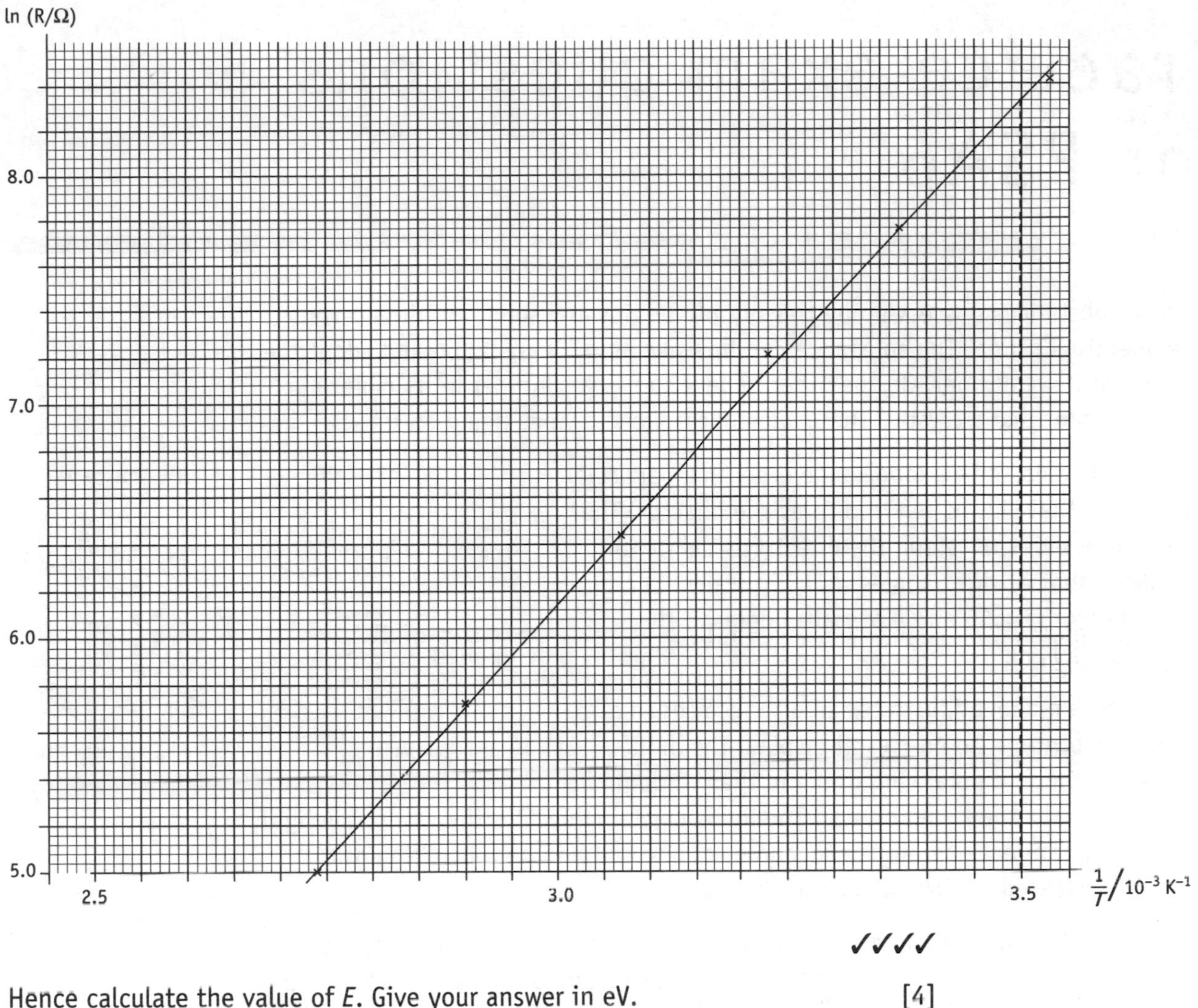

✓✓✓✓

Q Hence calculate the value of E. Give your answer in eV. [4]

A E is found from the gradient.

$$\ln(R) = \ln(A) + \frac{E}{2k} \times \frac{1}{T}$$

$$\text{gradient} = \frac{E}{2k}$$ ✓

$$\text{rise} = 8.32 - 5.00 = 3.32$$

$$\text{run} = (3.50 - 2.73) \times 10^{-3}\ \text{K}^{-1} = 0.77 \times 10^{-3}\ \text{K}^{-1}$$ ✓

Draw a large triangle that uses the 'best line', not the actual plotted points.

$$\text{gradient} = \frac{3.32}{0.77 \times 10^{-3}\ \text{K}^{-1}}$$

$$E = 2k \times \text{gradient} = \frac{2 \times 1.38 \times 10^{-23}\ \text{J K}^{-1} \times 3.32}{0.77 \times 10^{-3}\ \text{K}^{-1}} = 1.2 \times 10^{-19}\ \text{J}$$ ✓

$$= \frac{1.2 \times 10^{-19}\ \text{J}}{1.60 \times 10^{-19}\ \text{J eV}^{-1}} = 0.74\ \text{eV}$$ ✓

There is some uncertainty in the data and your answer will depend on exactly where you drew your 'best line' and your triangle, so the examiners' mark scheme allows for a range of values to gain full marks.

Practice exam questions for PSA6

1 Some substances possess properties of both liquids and solids within a certain temperature range. The substance can flow like an ordinary liquid and adopt the shape of its container. On the other hand it shows some order in its molecular arrangement like a crystal. These substances are known as liquid crystals.

Liquid crystals show a long-range order in a particular direction that will be referred to as the director. The refractive index, *n,* along the director is a different value to that at right angles to it. Typically n along the director is 1.65 and n at right angles to the director is 1.50.

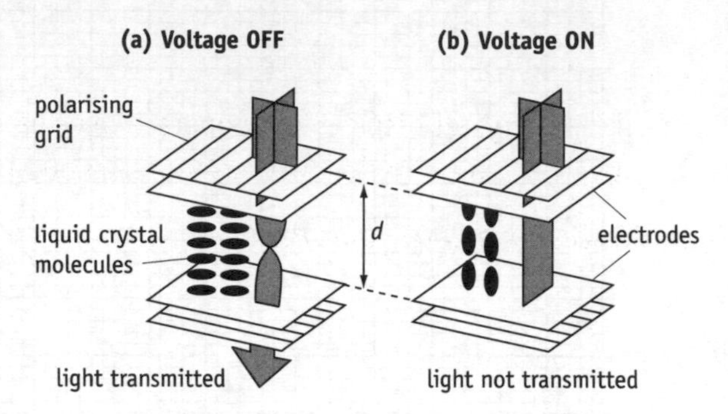

The diagram above illustrates an arrangement in a simple prototype liquid crystal display. The light passes through a polarising grid and then through an electrode into the liquid crystal. The liquid crystal molecules then twist the light so that it can leave through a second electrode and another polarising grid, which is crossed at right angles to the first.

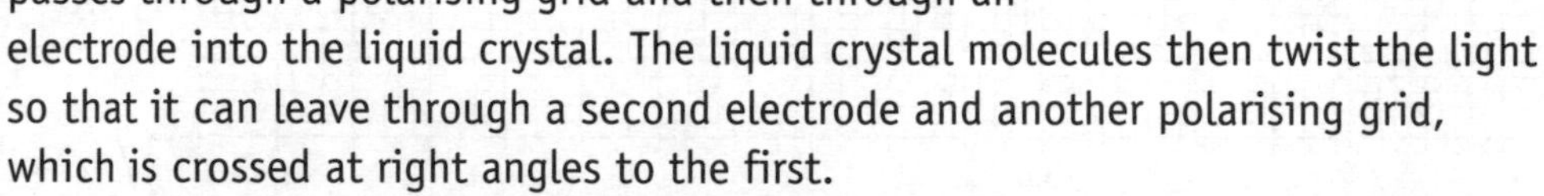

If the rotation of the plane of polarisation is 90° then the light is transmitted out of the bottom grid. This can be achieved using an appropriate thickness, *d,* of liquid crystal given by

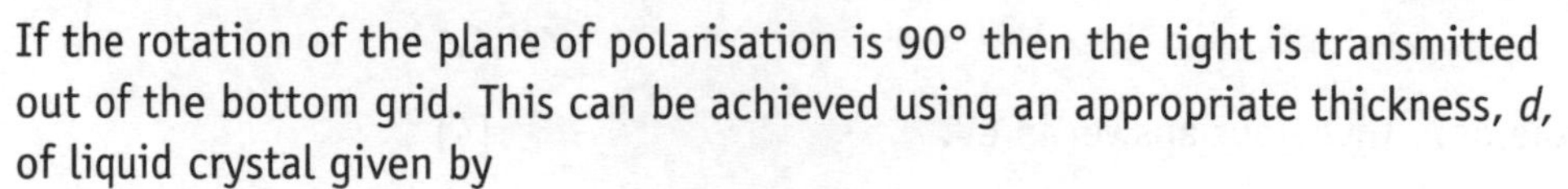

$$d = \frac{1.05 \text{ mm}}{\Delta n}$$

where Δn is the difference between the two refractive indices within the liquid crystal.

If a small voltage is now placed across the two electrodes, the resulting electric field alters the orientation of the director such that the effective value of Δn decreases. The rotation of the plane of polarisation is significantly reduced and the light is no longer transmitted.

Liquid crystals are used in calculator, wristwatch and laptop computer displays. A display usually consists of several segments each controlled by an independent voltage. Light is allowed through the liquid crystal segment unless the voltage switches it off.

One of the problems with liquid crystal displays is their turn-on time, T_{on}. This is the time it takes for the crystal to respond to the voltage. T_{on} is dependent on a number of factors and is proportional to the viscosity of the liquid crystal substance. If T_{on} is about 10 ms then for a display being scanned at 25 Hz, without multiplexing, only four parts of the display can be addressed in one complete scan.

a Briefly explain how scientists might investigate the order (or lack of it) in a liquid crystal. [2]

b Calculate the speed of light along the director of the liquid crystal. [2]

c What is meant by **plane polarised** light? [1]

d Describe an experiment that could be used to check whether the plane of polarisation had been rotated by a sugar solution. [4]

e Calculate the thickness d of liquid crystal required for a 90° rotation of the plane of polarisation. [2]

f Calculate the electric field strength across the liquid crystal for a voltage of 1.5 V. State any assumption made. [3]

g Explain, with a suitable calculation, why, without multiplexing, only four parts of the display can be addressed within each scan. [4]

h The problem discussed in **g** can be reduced by the use of **multiplexing**. Explain the meaning of the word **multiplexing** and describe one way in which multiplexing might be achieved. [4]

i Liquid crystal displays only operate over narrow temperature ranges. Give two reasons why they will not operate outside this range. [2]

From Edexcel June 2001, Unit PSA6ii [Total: 24]

2 It is suggested that the turn-on time, T_{on}, for a liquid crystal display is given by the equation

$$T_{on} = \frac{k\eta d^2}{V^2}$$

where η is the viscosity, V the voltage applied, d the thickness of crystal, and k is a constant.

Data showing how the turn-on time T_{on} depends on the voltage V are provided in the table below.

	turn-on time T_{on}/ms	voltage V/V	
	5	2.01	
	10	1.42	
	15	1.16	
	20	1.00	
	27	0.86	

a Plot a suitable graph to test the relationship suggested between T_{on} and V. Record the results of any calculations that you perform by adding to the table above. [7]

b Discuss whether or not your graph confirms the suggested relationship between T_{on} and V. [2]

c Use your graph to calculate a value of the constant k. The viscosity of the liquid crystal substance is 0.072 Pa s and its thickness is 6.0 μm. [4]

d A second experiment is to be carried out to test the relationship between the turn-on time T_{on} and the thickness d. Suggest a possible pair of axes to obtain a straight-line graph from measurements of T_{on} and d.

List all the possible variables that would need to be kept constant. [3]

From Edexcel June 2001, Unit PSA6ii [Total: 16]

Answers to quick check questions

Unit PSA4

Momentum conservation

Page 3

1 m = 100 g = 0.100 kg, $p = mv$ = 0.100 kg × 25 m s^{-1} = 2.5 kg m s^{-1}.

2 $F = ma$. A force of 1 N accelerates a mass of 1 kg by 1 m s^{-2} so 1 N = 1 kg m s^{-2}. Therefore 1 N s = 1 kg m s^{-2} × 1 s = 1 kg m s^{-1}.

3 Mass of truck = m = 4.0 t;
mass of locomotive = M = 32 t

Initial speed of truck = u;
initial speed of locomotive = U = 0

Speed of wagon and locomotive after collision = v = 4.0 m s^{-1}

It is helpful to choose symbols for the quantities you are using and list their values.

Momentum before collision = momentum after collision

$mu + MU = (M + m)v$

$$U = 0 \quad \text{so} \quad u = \frac{(M+m)v}{m} = \frac{(36\ \text{t} \times 4.0\ \text{m s}^{-1})}{4.0\ \text{t}}$$

= 36 m s^{-1}

4 a See figure. The combined momentum has magnitude 1200 kg m s^{-1} in a direction 35° to the original motion of the first player.

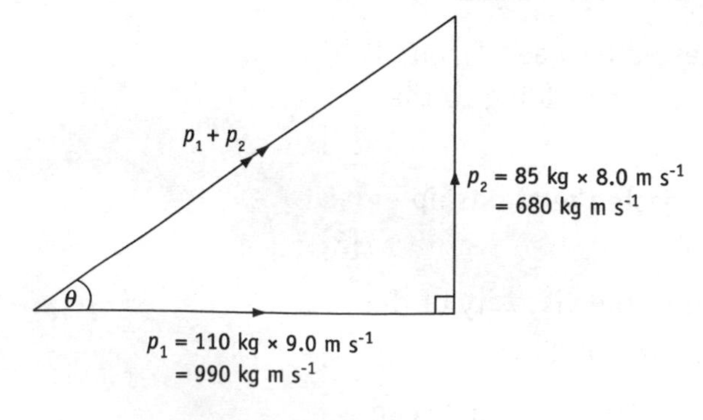

b Combined mass m = 195 kg so speed

$$v = \frac{1200\ \text{kg m s}^{-1}}{195\ \text{kg}} = 6.2\ \text{m s}^{-1}$$

5 No, because a falling object is *not* a system on which no external force is acting. It is interacting with the Earth via the force of gravity.

Momentum and force

Page 4

1 a Initial momentum $p_1 = mv_1$
= +20 × 10^3 kg × 15 m s^{-1} = +3.0 × 10^5 kg m s^{-1}.

Include the sign because it gives information about direction.

b Final momentum p_2 = 0 so $\Delta p = p_2 - p_1$
= −3.0 × 10^5 kg m s^{-1}

$$F = \frac{\Delta p}{\Delta t} = \frac{-3.0 \times 10^5\ \text{kg m s}^{-1}}{10\ \text{s}} = -3.0 \times 10^4\ \text{N}$$

The negative sign shows that the force acts in the opposite direction to the wagon's motion.

2 Force = rate of change of momentum
= 0.020 kg s^{-1} × 35 m s^{-1} = 0.70 kg m s^{-2} = 0.70 N

Momentum and kinetic energy

Page 5

1 From $p = mv$, we have $v = p/m$. Then $E_k = \frac{1}{2}mv^2$
$= \frac{1}{2}m(p/m)^2 = mp^2/2m^2 = p^2/2m$,
so $p^2 = 2mE_k$ and $p = \sqrt{(2mE_k)}$.

2 **a** A, **b** B, **c** B, **d** D, **e** C, **f** A

3 a They have the *same momentum*, so the smaller mass (the alpha particle) has the greater kinetic energy.

b
$$\frac{E_k\ \text{alpha}}{E_k\ \text{thorium}} = \frac{(p^2 / 2m\ \text{alpha})}{(p^2 / 2m\ \text{thorium})}$$

$$= \frac{m\ \text{thorium}}{m\ \text{alpha}} = \frac{234\ \text{u}}{4\ \text{u}} = 58.5$$

The alpha has 58.5 times the kinetic energy of the thorium nucleus.

Energy and momentum in collisions

Page 7

1 Momentum after = momentum before

$(M + m)\ v = mu + MU$

where m = mass of car = 800 kg

M = mass of truck = 2000 kg

u = initial velocity of car = +15.0 m s^{-1}

U = initial velocity of truck = −5.0 m s^{-1}

v = final velocity.

We have taken the car's initial direction as positive so U is negative.

Momentum:

$$v = \frac{(mu + MU)}{(M + m)}$$

$$= \frac{(800\ \text{kg} \times 15.0\ \text{m s}^{-1} - 2000\ \text{kg} \times 5.0\ \text{m s}^{-1})}{2800\ \text{kg}}$$

$$= \frac{2000\ \text{kg m s}^{-1}}{2800\ \text{kg}} = +0.71\ \text{m s}^{-1}$$

The positive sign shows that they move in the same direction as the car's initial motion.

Kinetic energy:

E_k before collision $= \frac{1}{2} mu^2 + \frac{1}{2} MU^2$

$= \frac{1}{2} \times 800\ \text{kg} \times (15.0\ \text{m s}^{-1})^2 +$

$\frac{1}{2} \times 2000\ \text{kg} \times (5.0\ \text{m s}^{-1})^2$

$= 1.15 \times 10^5$ J

E_k after collision $= \frac{1}{2} (M + m)v^2$

$= \frac{1}{2} \times 2800\ \text{kg} \times (0.71\ \text{m s}^{-1})^2$

$= 706$ J

E_k after/E_k before = 6.1×10^{-3} = 0.061% so almost 100% of the initial kinetic energy has been lost.

Circular motion

Page 11

1 2π, π, $\pi/2$, $\pi/3$, $\pi/4$

2 57.3°, 14.3°, 180°, 45°

3 **a** $f = 720\ \text{min}^{-1} = 720/60\ \text{rev s}^{-1} = 12$ Hz

$\omega = 2\pi f = 75.4\ \text{rad s}^{-1}$

b $v = r\omega = 3.0 \times 10^{-3}\ \text{m} \times 75.4\ \text{s}^{-1} = 0.23\ \text{m s}^{-1}$

4 $a = v^2/r$ so $v^2 = ar = 1.6\ \text{m s}^{-2} \times 1.8 \times 10^6$ m $= 2.88 \times 10^6\ \text{m}^2\ \text{s}^{-2}$ and $v = 1.7 \times 10^3\ \text{m s}^{-1}$

5 Distance travelled once round circle = $2\pi r$ $= 2\pi \times 1.2$ m.

One revolution takes 1.4 s, so

speed $v = 2\pi \times 1.2\ \text{m}/1.4\ \text{s} = 5.39\ \text{m s}^{-1}$

Or use other routes to get the same result, e.g.

$T = 2\pi/\omega$ so $\omega = 2\pi/T$ and so $v = r\omega = 2\pi r/T$.

Assuming we can approximate the hammer to a point mass and ignore the mass of the athlete's arms:

$$F = \frac{mv^2}{r} = \frac{4.0\ \text{kg} \times (5.39\ \text{m s}^{-1})^2}{1.2\ \text{m}} = 97\ \text{N}$$

6 To make an object move in a circle, it must be given a centripetal acceleration. This acceleration must be provided by a centripetal force, i.e. one that acts towards the centre of the circle. When something is whirled on a string, the tension in the string (the person pulling on the string) provides the necessary force. If you let go of the string the force stops acting. As there is no net force on the object it keeps moving in a straight line so it travels along the tangent to the circle.

When you are asked to 'explain' you need to say what *happens and* why.

Digital signals

Page 13

1 Number of levels = 2^8 = 256.

Sending signals

Page 15

1 In each second there are 44×10^3 samples. Each sample has 8 bits, so each signal has $8 \times 44 \times 10^3$ bits per second = 3.52×10^5 bits per second. Total number of signals that can pass along the cable = $\frac{10^9\ \text{Hz}}{3.52 \times 10^5\ \text{s}^{-1}} = 2.8 \times 10^3$.

2 **a** A, **b** A, **c** C, **d** B, **e** C, **f** C

Capacitors

Page 17

1 **a** $C = \frac{Q}{V} = \frac{40 \times 10^{-6}\ \text{C}}{5.0\ \text{V}} = 8.0 \times 10^{-6}\ \text{F}\ (= 8.0\ \mu\text{F})$

b $Q = CV = 8.0 \times 10^{-6}\ \text{F} \times 20\ \text{V} = 160\ \mu\text{C}$

2 As the charge on the capacitor decreases, the pd across it also decreases. Since V is directly proportional to Q, the graphs have the same shape. The pd across the resistor is the same as the pd across the capacitor. As the pd decreases so does the current I. As I is directly proportional to V, the current graph has the same shape as the other two.

3 **a** Using $Q = CV$ to substitute for Q in $W = \frac{1}{2} QV$:

$W = \frac{1}{2} CV \times V = \frac{1}{2} CV^2$

b Using $V = Q/C$ to substitute for V:

$W = \frac{1}{2} Q \times (Q/C) = \frac{1}{2} Q^2/C$

4 a $W = \frac{1}{2}QV = \frac{1}{2}CV^2$

so $C = \frac{2W}{V^2} = \frac{2 \times 10 \times 10^{-3}\ \text{J}}{(100\ \text{V})^2} = 2.0 \times 10^{-6}$ F (= 2.0 μF)

b Since $W = \frac{1}{2}CV^2$, doubling the pd will multiply the energy by 4 so the stored energy is now 40 mJ.

Exponential decay

Page 19

1 $\tau = RC = 5 \times 10^6\ \Omega \times 20 \times 10^{-6}\ \text{F} = 100$ s

2 The ratio of the time constants is

$$\frac{40 \times 10^{-6}\ \text{F} \times 2.5 \times 10^6\ \Omega}{1000 \times 10^{-6}\ \text{F} \times 100 \times 10^3\ \Omega} = \frac{100}{100}$$

so the time constants are both the same.

3 See figure. Both have the same initial charge Q_0 because the capacitance and pd are the same. Both decay exponentially. The second discharge (curve 2) has twice the time constant of the first (curve 1) because the resistance is doubled. The charge therefore takes twice as long to fall to 37% of its initial value.

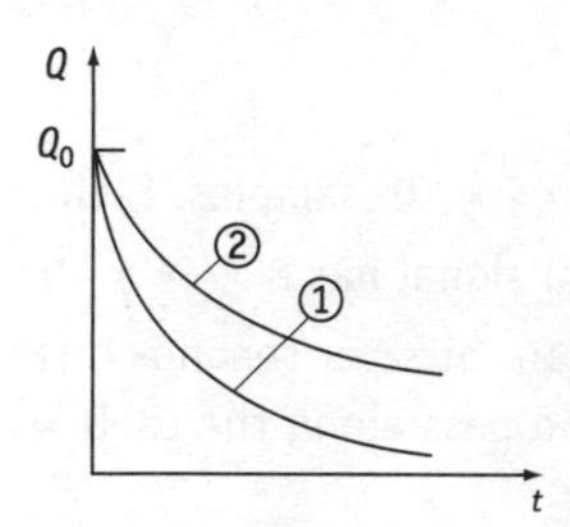

4 a $Q = CV = 20 \times 10^{-12}\ \text{F} \times 100\ \text{V} = 2.0 \times 10^{-9}$ C

b $\tau = RC = 500 \times 10^6\ \Omega \times 20 \times 10^{-12}\ \text{F} = 1.0 \times 10^{-2}$ s (= 0.010 s)

c $t/RC = 0.015\ \text{s}/0.010\ \text{s} = 1.5$

$Q = Q_0\, e^{-t/RC} = 2.0 \times 10^{-9}\ \text{C} \times e^{-1.5} = 4.5 \times 10^{-10}$ C

5 $I/I_0 = e^{-\mu x}$ so $\ln (I/I_0) = -\mu x$

When $x = 3 \times 10^{-17}$ m, $I/I_0 = 10\% = 0.10$

$-\mu x = -\ln (0.10) = -2.3$

$\mu = 2.3/(3 \times 10^{17}\ \text{m}) = 7.7 \times 10^{-18}\ \text{m}^{-1}$

Exponential relationships and log graphs

Page 21

1 See table.

Data for a cable

Distance x/km	**Signal voltage V/V**	**ln (V)**
0	6.00	1.79
100	4.67	1.54
200	3.63	1.29
300	2.83	1.04
400	2.21	0.79
500	1.72	0.54
600	1.34	0.29

From the figure, the signal voltage halves after about 275 km and halves again after a further 275 km so the attenuation is exponential. The signal voltage falls to 1/e of its initial value after a distance of 400 km. (0.37 × 6.00 V = 2.22 V).

So $\mu = \frac{1}{400 \times 10^3\ \text{m}} = 2.5 \times 10^{-6}\ \text{m}^{-1}$.

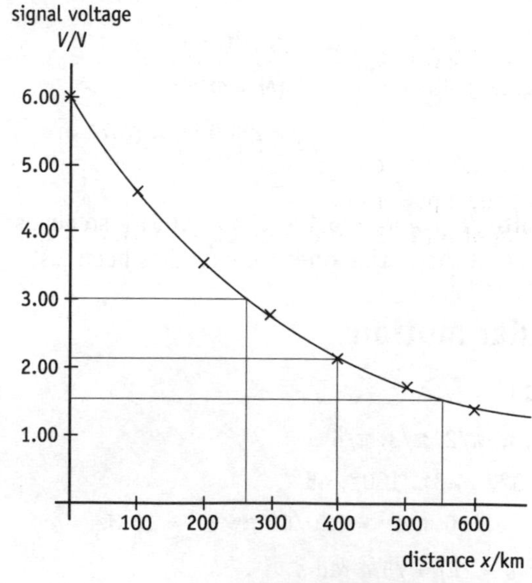

From the second figure (opposite page), the log-linear graph of ln (V) against x is a straight line, again confirming exponential attenuation.

Gradient $\mu = \frac{1.50}{600 \times 10^3\ \text{m}} = 2.5 \times 10^{-6}\ \text{m}^{-1}$

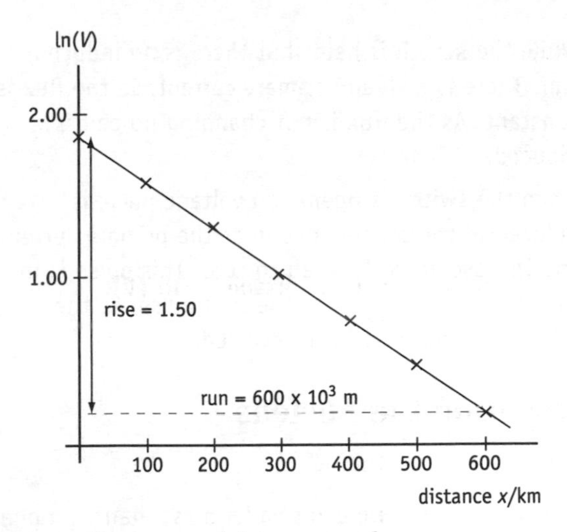

Power laws and log graphs

Page 23

1 See table. We have used base 10 logs.

Data for Solar System planets

planet	$r/10^6$ km	T/years	log (r)	log (T)
Mercury	57.9	0.241	1.76	-0.62
Venus	108.2	0.612	2.03	-0.21
Earth	149.6	1.00	2.17	0.00
Mars	227.9	1.52	2.36	0.18
Jupiter	778.3	11.9	2.89	1.08
Saturn	1427	29.5	3.15	1.47
Uranus	2870	84.0	3.46	1.92
Neptune	4497	165	3.65	2.22
Pluto	5900	247	3.77	2.39

The graph in the figure is a straight line so the data do obey a power law.
Gradient = (2.73 + 0.80)/(4.00 − 1.63) = 1.49.

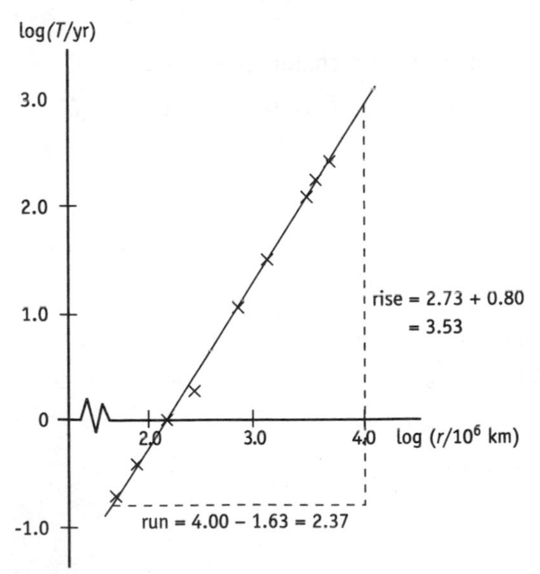

So from this graph p = 1.49. This is very close to 1.5, which is a much more likely value. (You probably got a slightly different value but it should still be very close to 1.5.)

Electric forces and fields

Page 25

1 $F = QE = eE = 1.60 \times 10^{-19}\ \text{C} \times 1.5 \times 10^{3}\ \text{N C}^{-1}$
$= 2.4 \times 10^{-16}\ \text{N}$

2 $E = \dfrac{V}{d} = \dfrac{120\ \text{V}}{0.15\ \text{m}} = 800\ \text{V m}^{-1}$

3 $F = QE = \dfrac{QV}{d} = \dfrac{3 \times 10^{-6}\ \text{C} \times 45\ \text{V}}{0.25\ \text{m}} = 5.4 \times 10^{-4}\ \text{N}$

4 $F = \dfrac{kQ_1Q_2}{r^2} = \dfrac{8.99 \times 10^{9}\ \text{N m}^2\ \text{C}^{-2} \times (1.60 \times 10^{-19}\ \text{C})^2}{(2.0 \times 10^{-15}\ \text{m})^2}$
$= 58\ \text{N}$

5 $r = 20\ \text{mm} + 10\ \text{mm} = 30\ \text{mm} = 30 \times 10^{-3}\ \text{m}$

$E = \dfrac{kQ}{r^2} = \dfrac{8.99 \times 10^{9}\ \text{N m}^2\ \text{C}^{-2} \times 70 \times 10^{-6}\ \text{C}}{(30 \times 10^{-3}\ \text{m})^2}$

$= 7.0 \times 10^{8}\ \text{N C}^{-1}$

Particles in electric fields

Page 27

1 a Energy = 5.0 keV

b $5.0\ \text{keV} = 5.0 \times 10^{3} \times 1.60 \times 10^{-19}\ \text{J}$
$= 8.0 \times 10^{-16}\ \text{J}$

2 $\text{Energy} = \dfrac{6.21 \times 10^{-21}\ \text{J}}{1.60 \times 10^{-19}\ \text{J eV}^{-1}} = 3.88 \times 10^{-2}\ \text{eV}$

3 $eV = \frac{1}{2} mv^2$ so $V = mv^2/2e$

$V = \dfrac{1.67 \times 10^{-27}\ \text{kg} \times (1.2 \times 10^{7}\ \text{m s}^{-1})^2}{2 \times 1.60 \times 10^{-19}\ \text{C}}$

$= 7.5 \times 10^{5}\ \text{V}$ (= 0.75 MV)

Magnetic forces and fields

Page 29

1 a sin 90° = 1, so $F = BI\ell$
$= 80 \times 10^{-3}\ \text{T} \times 5.0\ \text{A} \times 2.0\ \text{m} = 0.80\ \text{N}$

b $F = BI\ell \sin 60° = 0.80\ \text{N} \times \sin 60° = 0.69\ \text{N}$

c $F = 0$ because $\sin 0° = 0$

2 a sin 90° = 1, so $F = Bqv$
$= 0.20\ \text{T} \times 1.60 \times 10^{-19}\ \text{C} \times 1.0 \times 10^{7}\ \text{m s}^{-1}$
$= 3.2 \times 10^{-13}\ \text{N}$

b $F = Bqv \sin 40° = 3.2 \times 10^{-13}\ \text{N} \times \sin 40°$
$= 2.1 \times 10^{-13}\ \text{N}$

3 $r = \dfrac{mv}{Bq} = \dfrac{1.67 \times 10^{-27}\ \text{kg} \times 2.5 \times 10^{6}\ \text{m s}^{-1}}{1.2\ \text{T} \times 1.60 \times 10^{-19}\ \text{C}}$
$= 2.2 \times 10^{-2}\ \text{m}$

4 $p = Bqr = 0.75\ \text{T} \times 1.60 \times 10^{-19}\ \text{C} \times 4.5 \times 10^{-2}\ \text{m}$
$= 5.4 \times 10^{-21}\ \text{kg m s}^{-1}$

If the charge were $2e$, the momentum would be twice that calculated above.

Particles in magnetic and electric fields

Page 31

1 **a** A, **b** B

Electromagnetic induction

Page 33

1 $\Phi = BA$, and $A = \pi r^2$

$\Phi = 250 \times 10^{-3}\ \text{T} \times \pi \times (6.0 \times 10^{-2}\ \text{m})^2$
$= 2.83 \times 10^{-3}\ \text{Wb}$

2 flux linkage $N\Phi = 40 \times 2.83 \times 10^{-3}\ \text{Wb} = 0.113\ \text{Wb}$

3 $\mathscr{E} = -\text{d}(N\Phi)/\text{d}t = 0.113\ \text{Wb}/0.20\ \text{s} = 0.57\ \text{V}$

Transformers

Page 35

1 **a** $N_s/N_p = V_s/V_p = 24\ \text{V}/240\ \text{V} = 0.10$

b $I_s = V_s/R = 24\ \text{V}/8\ \Omega = 3\ \text{A}$

c $I_p/I_S = N_s/N_p = 0.10$ so $I_p = 0.10 I_s = 0.3\ \text{A}$

2 **a** From the graph, peak $V_p = 340$ V, peak $V_s = 1700$ V
$N_s/N_p = V_s/V_p = 1700/340 = 5$

b When V_p is passing through zero it is changing rapidly. The current and its associated magnetic field are also changing rapidly. This means that the flux linking the secondary coil must be changing rapidly, so a large emf V_s is induced. At the instant V_p is passing through a peak, its value is not changing (the graph is horizontal). The current and flux are also not changing. As the flux in the secondary coil is not changing, no emf can be induced so V_s is zero.

3 See figure. As the switch is closed a voltage pulse is induced in the secondary coil. As the primary current increases from zero the flux in the coil changes, inducing a secondary emf.

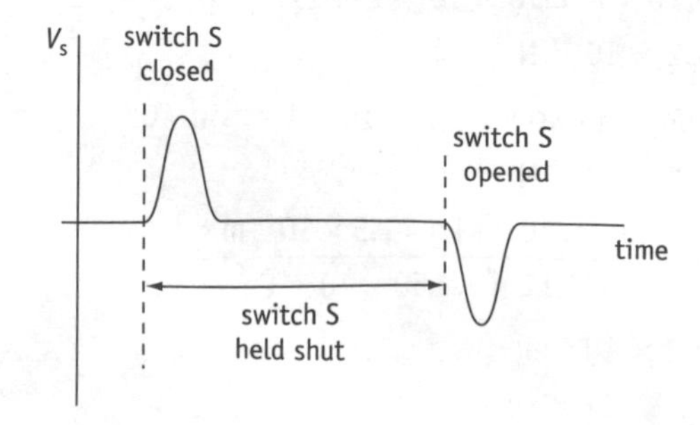

While the switch is held shut there is no induced emf. There is a steady primary current, so the flux is constant. As the flux is not changing no emf is induced.

When the switch is opened, a voltage pulse is induced in the secondary coil as the primary current and its associated flux fall to zero. This pulse is in the opposite direction to the first pulse, as the original change has been reversed.

Discovering the nucleus

Page 37

1 The electron has a much smaller mass than an alpha particle. If a fast-moving alpha particle collided with an electron, the electron would be knocked aside and the alpha would continue more or less unaffected.

2 The alphas would undergo many small deflections as they passed through the positive and negative charge. They would not go straight through, nor would they bounce back at large angles.

3 $p = \dfrac{h}{\lambda} = \dfrac{6.63 \times 10^{-34}\ \text{J s}}{8.6 \times 10^{-11}\ \text{m}} = 7.7 \times 10^{-24}\ \text{kg m s}^{-1}$

4 $E_k = eV$ so the electrons' kinetic energy E_k is multiplied by 4.

$E_k = p^2/2m$, $p = \sqrt{(2mE_k)} = \sqrt{(2meV)}$ so the momentum is doubled to $1.5 \times 10^{-23}\ \text{kg m s}^{-1}$.
$\lambda = h/p$ so the wavelength is halved to 4.3×10^{-11} m.

Particles

Page 39

1 It is a meson, with charge $\frac{-1}{3}e + (\frac{-2}{3}e) = -e$

2 ${}^{7}_{3}\text{Li} + {}^{1}_{1}\text{p} \rightarrow {}^{7}_{4}\text{Be} + {}^{1}_{0}\text{n}$, so Be has charge +4 and 7 neutrons.

3 $\pi^+ \rightarrow \overline{\mu} + \nu_\mu$

Particle interactions

Page 41

1 $\Delta m = 2 \times m_p$

Each photon has energy $E = m_p c^2$

$E = hf$ and $c = f\lambda$ so $E = \dfrac{hc}{\lambda}$

hence $\lambda = \dfrac{hc}{E} = \dfrac{hc}{m_p c^2} = \dfrac{h}{m_p c}$

$= \dfrac{6.63 \times 10^{-34}\ \text{J s}}{1.67 \times 10^{-27}\ \text{kg} \times 3.00 \times 10^{8}\ \text{m s}^{-1}}$

$= 1.32 \times 10^{-15}$ m

2 $m = 180$ GeV/c^2

$$= \frac{180 \times 10^9 \times 1.60 \times 10^{-19}\ \text{J}}{(3.00 \times 10^8\ \text{m s}^{-1})^2}$$

$= 3.2 \times 10^{-25}$ kg

3 Mass 1.3 GeV/c^2 and charge $-\frac{2}{3}e$.

4 $\Delta m = 2 \times 4.3$ GeV/c^2

$\Delta E = c^2 \Delta m = 2 \times 4.3$ GeV

Energy of each photon $E = 4.3$ GeV

$= 4.3 \times 1.60 \times 10^9 \times 10^{-19}$ J

$E = hf$ and $c = f\lambda$ so $\lambda = hc/E$

$$\lambda = \frac{6.63 \times 10^{-34}\ \text{J s} \times 3.00 \times 10^8\ \text{m s}^{-1}}{4.3 \times 1.60 \times 10^9 \times 10^{-19}\ \text{J}}$$

$= 2.89 \times 10^{-16}$ m

Unit PSA5

Simple harmonic motion

Page 47

1 a $\omega = 2\pi/T = 2\pi/3.0\ \text{s} = 2.1$ rad s^{-1}

b Max $v = A\omega = 0.20\ \text{m} \times 2.1\ \text{rad s}^{-1} = 0.42\ \text{m s}^{-1}$

c Max $a = A\omega^2 = 0.20\ \text{m} \times (2.1\ \text{rad s}^{-1})^2 = 0.88\ \text{m s}^{-2}$

2 $\omega t = 7.4\ \text{rad s}^{-1} \times 2.0\ \text{s} = 14.8$ rad

$x = A \cos(\omega t) = 2.5 \times 10^{-3}\ \text{m} \times \cos(14.8\ \text{rad})$

$= 2.5 \times 10^{-3}\ \text{m} \times (-0.615) = -1.5 \times 10^{-3}$ m

$v = -A\omega \sin(\omega t)$

$= -2.5 \times 10^{-3}\ \text{m} \times 7.4\ \text{rad s}^{-1} \times \sin(14.8\ \text{rad})$

$= -2.5 \times 10^{-3}\ \text{m} \times 7.4\ \text{rad s}^{-1} \times 0.788$

$= -1.5 \times 10^{-2}\ \text{m s}^{-1}$

$a = -A\omega^2 \cos(\omega t) = -\omega^2 x$

$= -(7.4\ \text{rad s}^{-1})^2 \times (-1.5 \times 10^{-3}\ \text{m})$

$= 8.2 \times 10^{-2}\ \text{m s}^{-2}$

3 a No. While the person is in the air they experience a constant downward force due to gravity, and when they contact the ground they experience a large upward force for a short time.

b Probably yes. Provided the elastic obeys Hooke's law, the force will be proportional to the displacement and in the opposite direction as required for SHM.

c No. While they are in contact with the trampoline the force might satisfy the conditions for SHM, but in the air the force is constant and does not depend on displacement.

Energy damping and forced oscillations

Page 51

1 a Max E_k occurs when $\sin^2 \omega t = 1$ so $E_k = \frac{1}{2} mA^2\omega^2$

b Max E_p occurs when $\cos^2 \omega t = 1$ so $E_p = \frac{1}{2} kA^2$

2 a $\omega = 2\pi/T = 2\pi/3.0\ \text{s} = 2.1$ rad s^{-1}

Max $E_k = \frac{1}{2} mA^2\omega^2$

$= \frac{1}{2} \times 0.10\ \text{kg} \times (0.20\ \text{m} \times 2.1\ \text{rad s}^{-1})^2$

$= 8.8 \times 10^{-3}$ J

b There are at least two ways to approach this.

Either (i) use $\omega^2 = k/m$:

$k = m\omega^2 = 0.10\ \text{kg} \times (2.1\ \text{rad s}^{-1})^2 = 0.44\ \text{N m}^{-1}$

or (ii) use energy:

Max E_k = max $E_p = 8.8 \times 10^{-3}$ J

Max $E_p = \frac{1}{2} kA^2$ so

$$k = \frac{2E_{max}}{A^2} = \frac{2 \times 8.8 \times 10^{-3}\ \text{J}}{(0.20\ \text{m})^2} = 0.44\ \text{N m}^{-1}$$

c When $E_k = E_p$, each is at half the maximum value, so

$E_k = E_p = \frac{1}{2} \times 8.8 \times 10^{-3}\ \text{J} = 4.4 \times 10^{-3}$ J

Using $E_p = \frac{1}{2} kx^2$

$$x^2 = \frac{2E_p}{k} = \frac{2 \times 4.4 \times 10^{-3}\ \text{J}}{0.44\ \text{N m}^{-1}} = 0.02\ \text{m}^2$$

so $x = 0.14$ m

3 a At a certain speed, the vibrations caused by the car's engine and /or wheels match the mirror's own natural frequency. Resonance occurs: energy is transferred to the mirror and oscillations build up to large amplitude.

b Adding Blu-tack will change the mirror's natural frequency (adding mass will reduce the frequency) and so will change the frequency at which it resonates. But it is unlikely to damp the motion, so the mirror will still shake noticeably – the shaking will now occur at a different speed.

Seismic and sound waves

Page 53

1 $v = (E/\rho)^{1/2}$ so

$E = \rho v^2 = 7700\ \text{kg m}^{-3} \times (5100\ \text{m s}^{-1})^2$

$= 2.1 \times 10^{11}\ \text{N m}^{-2}$

2 The intensity is one-quarter its initial value.

$I = kA^2$ where A is amplitude and k is a constant.

Initially $A = A_0$ and $I = I_0$.

If $A = A_0/2$ then $I = k(A_0/2)^2 = kA_0^2/4 = I_0/4$.

Radioactive decay

Page 57

1 $^{16}_{7}\text{N} \rightarrow {}^{0}_{-1}\text{e} + {}^{16}_{8}\text{O}$

2 a $\lambda = \ln(2)/t_{1/2} = \ln(2)/7.1\text{ s} = 0.098\text{ s}^{-1}$

b Mass, m, of ^{16}N present is directly proportional to the number of ^{16}N nuclei so

$m = m_0\, e^{-\lambda t}$

Taking reciprocals then natural logs of both sides

$\ln(m_0/m) = \lambda t$

$t = \ln(5.0\ \mu\text{g}/1.0\ \mu\text{g})/0.098\text{ s}^{-1} = 16.4\text{ s}$

c Mass of ^{16}N nucleus $= 16 \times 1.67 \times 10^{-27}$ kg

Initial mass of ^{17}N $= 5.0\ \mu\text{g} = 5.0 \times 10^{-6}$ g

$= 5.0 \times 10^{-9}$ kg

N_0 = no. of nuclei in 5.0 g

$$= \frac{5.0 \times 10^{-9}\text{ kg}}{16 \times 1.67 \times 10^{-27}\text{ kg}} = 1.87 \times 10^{17}$$

$A_0 = \lambda N_0 = 0.098\text{ s}^{-1} \times 1.87 \times 10^{17} = 1.8 \times 10^{16}$ Bq

Nuclear processes

Page 59

1 Mass of nucleons $= 26m_p + 28m_n$

$\Delta m = 26 \times 1.00728\text{ u} + 28 \times 1.00867\text{ u} - 53.93962\text{ u}$

$= 0.4924$ u

$= 0.4924 \times 1.67 \times 10^{-27}\text{ kg} = 8.22 \times 10^{-28}$ kg

$\Delta E = c^2\Delta m = (3.00 \times 10^8\text{ m s}^{-1})^2 \times 8.22 \times 10^{-28}$ kg

$= 7.40 \times 10^{-11}$ J

$$= \frac{740 \times 10^{-11}}{1.60 \times 10^{-13}}\text{ MeV} = 463\text{ MeV}$$

Binding energy per nucleon of $^{54}_{26}\text{Fe} = 463\text{ MeV}/54$

$= 8.57$ MeV per nucleon

2 **a** A, **b** B, **c** C, **d** C, **e** D, **f** C, **g** A

3 a $4\,{}^{1}_{1}\text{H} \rightarrow {}^{4}_{2}\text{He} + 2\,{}^{0}_{1}\text{e} + 2\nu_e$

b Initial mass $= 4m_p = 4 \times 1.00728$ u

$\Delta m = 4 \times 1.00728\text{ u} - 4.00260\text{ u} = 0.02652$ u

$= 0.02652 \times 1.67 \times 10^{-27}$ kg

$\Delta E = c^2\Delta m = (3.00 \times 10^8\text{ m s}^{-1})^2 \times 0.02652 \times 1.67 \times 10^{-27}$ kg

$= 3.99 \times 10^{-12}$ J

This is the energy from fusion of four nuclei.

From 1 kg, energy $= 3.99 \times 10^{-12}\text{ J} \times 6.02 \times 10^{26}/4 = 6.00 \times 10^{14}$ J

Stars

Page 61

1 You could find the Sun's distance and luminosity from information given on the page and use $d^2 = L/4\pi F$. A neater way uses a little algebra:

$F = L/4\pi d^2$, so the value of Fd^2/L is the same for all stars.

$(Fd^2/L)_{dwarf} = (Fd^2/L)_{Sun}$

$$\text{so } (d_{dwarf}/d_{Sun})^2 = \frac{(L_{dwarf} / L_{Sun})}{(F_{dwarf} / F_{Sun})}$$

$$d^2_{dwarf} = \frac{(L_{dwarf} / L_{Sun})}{(F_{dwarf} / F_{Sun})} \times d^2_{Sun}$$

$$= \frac{10^{-3}}{(7.0 \times 10^{-13}\text{ W m}^{-2} / 1.4 \times 10^3\text{ W m}^{-2})} \times d^2_{Sun}$$

$= 2.0 \times 10^{12}\, d^2_{Sun}$

so $d_{dwarf} = 1.4 \times 10^6\, d_{Sun}$

2 L is about $3 \times 10^5\, L_{Sun}$. It is about half-way between 10^5 and 10^6 on the logarithmic scale.

Gravitation

Page 63

1 $$T^2 = \frac{4\pi^2 r^3}{GM} \quad \text{so} \quad M = \frac{4\pi^2 r^3}{GT^2}$$

$T = 27.3\text{ days} = 27.3 \times 24 \times 60 \times 60$ s

$= 2.36 \times 10^6$ s

$$M = \frac{4\pi^2 \times (3.8 \times 10^8\text{ m})^3}{(6.67 \times 10^{-11}\text{ N m}^2\text{ kg}^{-2}) \times (2.36 \times 10^6\text{ s})^2}$$

$= 5.83 \times 10^{24}$ kg

2 **a** B, **b** A, **c** C, **d** C, **e** A, **f** B

3 a $$g = \frac{GM}{r^2}$$

$$\text{so } r^2 = \frac{GM}{g}$$

$$= \frac{6.67 \times 10^{-11}\text{ N m}^2\text{ kg}^{-2} \times 5.83 \times 10^{24}\text{ kg}}{9.81\text{ N kg}^{-1}}$$

$r = 6.3 \times 10^6$ m

b Density $\rho = M/V$

Sphere has volume $V = 4\pi r^3/3$

$$\text{so } \rho = \frac{3M}{4\pi r^3} = \frac{3 \times 5.83 \times 10^{24}\text{ kg}}{4\pi \times (6.3 \times 10^6\text{ m})^3}$$

$= 5.57 \times 10^3$ kg m^{-3}

4 First do some algebra to get an expression for g in terms of r and ρ.

$g = GM/r^2$

$M = \rho V = 4\pi r^3\rho/3$

so $g = G \times 4\pi r^3\rho/3r^2 = 4\pi Gr\rho/3$

If density ρ is halved and radius r multiplied by 4, then g is multiplied by 2 so the answer is A.

Kinetic theory and gas laws

Page 67

1 The molecular kinetic energy increases as the temperature rises. While the water is boiling at a constant temperature (100 °C) there is no increase in kinetic energy but the potential energy of the molecules increases.

2 $\frac{1}{2} m\langle c^2 \rangle = 3kT/2$ so if T doubles then $\langle c^2 \rangle$ doubles.

To double the mean square speed, the mean speed must be multiplied by $\sqrt{2}$, i.e. by 1.4.

3 $pV = nRT$ so $n = pV/RT$

$1\ \text{m}^3 = 10^6\ \text{cm}^3$ so $V = 2.24 \times 10^7\ \text{cm}^3 = 22.4\ \text{m}^3$

$$n = \frac{1.0 \times 10^5\ \text{Pa} \times 22.4\ \text{m}^3}{8.31\ \text{J mol}^{-1}\ \text{K}^{-1} \times 295\ \text{K}} = 914\ \text{mol}$$

4 When the freezer door is open, the air temperature inside the freezer rises slightly. The inside and outside are at atmospheric pressure. When the door is closed, the freezer cools the air inside it, reducing its pressure. The pressure outside the freezer is now greater than the pressure inside, so there is a net inward force on the door which makes it hard to open. Air gradually leaks into the freezer so that its pressure rises to match the pressure outside, so the door becomes easier to open after a while.

The universe

Page 70

1 $\Delta\lambda = 512\ \text{nm} - 434\ \text{nm} = 78\ \text{nm}$

$$z = \frac{\Delta\lambda}{\lambda} = \frac{78\ \text{nm}}{434\ \text{nm}} = 0.180$$

$v = cz = 0.180 \times 3.00 \times 10^8\ \text{m s}^{-1}$

$= 5.39 \times 10^7\ \text{m s}^{-1} = 5.39 \times 10^4\ \text{km s}^{-1}$

$$d = \frac{v}{H_0} = \frac{5.39 \times 10^4\ \text{km s}^{-1}}{75\ \text{km s}^{-1}\ \text{Mpc}^{-1}} = 719\ \text{Mpc}$$

2 a $H_0 = 75\ \text{km s}^{-1}\ \text{Mpc}^{-1} = \dfrac{75\ \text{km s}^{-1}}{1\ \text{Mpc}}$

$$= \frac{75 \times 10^3\ \text{m s}^{-1}}{1 \times 10^6 \times 3.09 \times 10^{16}\ \text{m}} = 2.43 \times 10^{-18}\ \text{s}^{-1}$$

$$\text{Age} = \frac{1}{H_0} = \frac{1}{2.43 \times 10^{-18}\ \text{s}^{-1}} = 4.12 \times 10^{17}\ \text{s}$$

$= 1.3 \times 10^{10}$ years

b This is an estimate because it assumes uniform expansion. In practice the expansion rate of the universe will have changed over time due to the effects of the matter it contains.

3 a A closed universe is one in which there is sufficient matter eventually to stop the expansion and make the universe contract.

b When the expansion stops, distant galaxies will be observed with zero redshift. At such a time, galaxies will on average be much further apart that today. During contraction, blueshifts will be observed as galaxies are approaching and their light is observed at shorter wavelengths than it is emitted.

Answers to practice exam questions

Unit PSA4

1 Distance between images is proportional to speed. Before collision, object A moves 3.5 cm on diagram between images so 3.5 cm represents 5.0 m s^{-1}

So 1 cm represents $\frac{5.0}{3.5}$ m s^{-1}, i.e. 1.43 m s^{-1} ✓

After collision, distance between images of A = 6.5 cm/4 = 1.6 cm

Speed = 1.6 × 1.43 m s^{-1} = 2.3 m s^{-1} ✓

Distance can be measured to the nearest mm ✓

So the distance is 6.5 cm ± 0.1 cm

Uncertainty = 0.1/6.5 = 0.015 = 1.5% ✓

There is no one right answer, as it will depend on how precisely you think you can measure the separation. But you are asked for a reasoned *estimate, so you need to justify any values that you use.*

Kinetic energy of A before collision = $\frac{1}{2} m_A u_A^2$

$= \frac{1}{2} \times 10 \text{ kg} \times (5.0 \text{ m s}^{-1})^2 = 125 \text{ J}$ ✓

Total kinetic energy afterwards = $\frac{1}{2} m_A v_A^2 + \frac{1}{2} m_B v_B^2$

$= \frac{1}{2} \times 10 \text{ kg} \times (2.3 \text{ m s}^{-1})^2 + \frac{1}{2} \times 5.0 \text{ kg} \times (6.0 \text{ m s}^{-1})^2 = 116 \text{ J}$ ✓

There is a net loss of kinetic energy so the collision is inelastic. ✓

Suppose the uncertainty in each of the speeds is about 2%.

If A's initial speed is 2% less (4.9 m s^{-1}) its E_k is 120 J. If both speeds afterwards are 2% more, then the total E_k afterwards is 120 J. So an uncertainty of 2% could mean that the collision was in fact elastic. ✓

This is not the only possible answer. You would get a mark for any *sensible statements about uncertainty.*

Momentum is always conserved in *all* directions. Initially there is no momentum in the *y* direction, so the total momentum in this direction afterwards must also be zero. ✓

Object A momentum = $m_A v_A \sin 24° = 10 \text{ kg} \times 2.3 \text{ m s}^{-1} \times \sin 24° = 9.35 \text{ kg m s}^{-1}$ ✓

Object B momentum = $m_B v_B \sin 17° = 5.0 \text{ kg} \times 6.0 \text{ m s}^{-1} \times \sin 17° = 8.77 \text{ kg m s}^{-1}$ ✓

The difference in these two values must be due to uncertainties in measurement, as momentum cannot be lost. ✓

2 Speech contains a wide range of frequencies. It is intelligible using frequencies up to 3500 Hz. ✓

You would get a mark for any sensible comments.

The sampling frequency must be at least twice the highest frequency in the signal. ✓

2×3500 Hz = 7 kHz, so a sampling frequency of 8 kHz is OK. ✓

There 64 000 bits per second. These encode 8×10^3 samples. ✓

So no. of bits in each sample = 64 000/8000 = 8 ✓

Each fibre can have 2×10^9 bits per second.

Each phone call needs 64 000 bits per second.

So no. of calls along each fibre = $2 \times 10^9/64\,000$ = 31 250, which is close to 32 000. ✓

This is achieved by time division multiplexing, i.e. each sampled signal is chopped into packets, and packets from each signal in turn are sent along the fibre. ✓

This is more than you would need for a mark. But always try to include a little extra detail as you cannot always be sure exactly what the examiners are looking for.

3 Circumference = $2\pi r$ so $r = 27\text{ km}/2\pi = 4.3$ km ✓

The electrons have been accelerated through a pd of 20 GV (20×10^9 V).

Their kinetic energy is $20 \times 10^9 \times 1.60 \times 10^{-19}\text{ J} = 3.2 \times 10^{-9}$ J. ✓

Electrons are travelling close to the speed of light. They are no longer described by laws of classical physic and behave as though their mass were much greater then their normal mass. ✓

You would get a mark for any sensible comment referring to their very high speed.

$$a = \frac{v^2}{r} = \frac{(3.0 \times 10^8\text{ m s}^{-1})^2}{4.3 \times 10^3\text{ m}} = 2.1 \times 10^{13}\text{ m s}^{-2}$$ ✓✓

$F = ma = 3.6 \times 10^{-26}\text{ kg} \times 2.1 \times 10^{13}\text{ m s}^{-2} = 7.6 \times 10^{-13}$ N, which is close to the value given. ✓

$$F = Bqv \quad \text{so} \quad B = \frac{F}{qv}$$ ✓

$$B = \frac{7.6 \times 10^{-13}\text{ N}}{1.60 \times 10^{-19}\text{ C} \times 3.0 \times 10^8\text{ m s}^{-1}} = 0.016\text{ T}$$ ✓

The electrons' speed cannot increase beyond c. But their kinetic energy and momentum are increased (which effectively means their mass increases). ✓

Since $r = p/Bq$, $B = p/rq$ an increase in momentum p requires an increase in field B to maintain the same radius, r, of orbit. ✓

Look for opportunities to quote relevant equations as part of an explanation.

4 See figure.

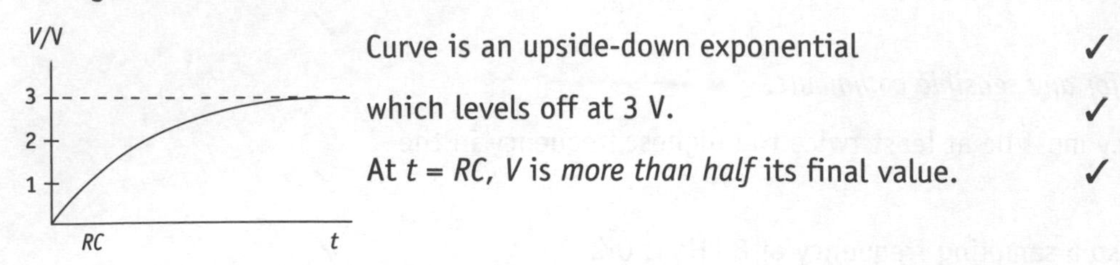

During discharge, V would fall to 0.37 × 3 V in this time. This curve is like an upside-down discharge curve.

Moving the diaphragm changes the capacitance C ✓

which changes the amount of charge that can be stored when the pd is 3 V. ✓

So as the diaphragm moves, charge must flow to/from the capacitor. ✓

As the charge is flowing, there is a current in R and a pd across R, which means there is an electrical signal. ✓

5 *You need to explain (1) why a voltage is generated (2) why the voltage is large.*

You would gain one mark for each correct statement that forms part of a logical explanation, up to a maximum of six marks. Any six of the following statements would gain full marks.

- When S is closed, there is a current in the primary coil.
- The current in the primary coil gives rise to a magnetic flux.
- The flux links both coils.
- When S is opened, the current and flux decrease.
- Changing the flux induces an emf in the secondary coil.
- Having a large number of turns N there is a large flux linkage $N\Phi$ in the secondary coil.
- The current and flux fall *rapidly* to zero.
- So the rate of change of flux linkage, $d(N\Phi)/dt$, is large
- which leads to a large induced emf $\mathscr{E} = -d(N\Phi)/dt$.

Look for opportunities to quote relevant equations in your explanation.

6 For a point charge, $E = Q/4\pi\varepsilon_0 r^2$

So $r^2 = \dfrac{Q}{4\pi\varepsilon_0 E}$ ✓

$= \dfrac{1.60 \times 10^{-19}\ \text{C}}{4\pi \times 8.85 \times 10^{-12}\ \text{F m}^{-1} \times 5 \times 10^{10}\ \text{V m}^{-1}} = 2.88 \times 10^{-20}\ \text{m}^2$ ✓

So $r = \sqrt{(2.88 \times 10^{-20}\ \text{m}^2)} = 1.7 \times 10^{-10}\ \text{m}$ ✓

This is about the size of an atom. So an electron in a (hydrogen) atom would be moving in a field of approximately this value. ✓

7 Energy equivalent to meson's mass, $\Delta E = c^2\Delta m = 135\ \text{MeV}/c^2 \times c^2$

$\Delta E = 135$ MeV ✓

$= 135 \times 10^6 \times 1.60 \times 10^{-19}\ \text{J} = 2.16 \times 10^{-11}\ \text{J}$ ✓

Energy of each photon, $E = 2.16 \times 10^{-11}\ \text{J}/2 = 1.08 \times 10^{-11}\ \text{J}$ ✓

$E = hf$ and $c = f\lambda$ so $\lambda = \dfrac{hc}{E}$ ✓

$$\lambda = \frac{6.63 \times 10^{-34}\ \text{J s} \times 3.00 \times 10^{8}\ \text{m s}^{-1}}{1.08 \times 10^{-11}\ \text{J}}$$

$= 1.84 \times 10^{-14}$ m ✓

Unit PSA5

1 Natural frequency (or fundamental frequency) ✓

'Resonant frequency' would not score a mark as the vibrations here are free not forced.

Resonance occurs when applied frequency equals the natural frequency ✓

which allows maximum energy transfer ✓

and hence large amplitude. ✓

You would score marks for any two of these points.

See the figure. ✓✓✓✓

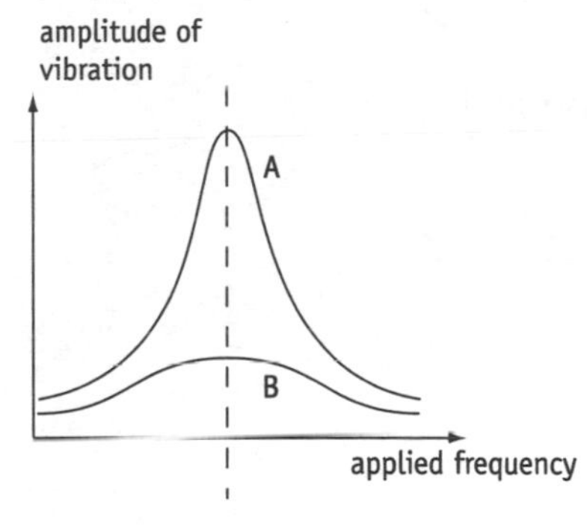

To score a mark curve A must be a reasonable shape with a clear peak. The entire curve B must lie underneath curve A and the peak must be broader.

Good suspension would provide damping ✓

so that resonant oscillations would not have large amplitude. ✓

2 See figure on page 98. Marks are awarded for:

Sound totally removed

- secondary speaker – in antiphase, same amplitude ✓
- error microphone – straight line (no signal). ✓

Sound significantly removed

- secondary speaker – in antiphase, different amplitude ✓
- error microphone – weak signal. ✓

	sound totally removed	sound significantly reduced
reference microphone		
secondary loudspeaker		
error microphone		

Any *two* of the following points: ✓✓

- The error microphone detects the resultant signal
- and provides a feedback signal to adjust the amplification
- so that the loudspeaker output is changed
- if the resultant signal is not zero.

3 Nuclei must be close together/must collide frequently. ✓

Nuclei must have large kinetic energy/high speed. ✓

(i) High density requires high pressure (particularly if temperature is also high). ✓

(ii) High temperature leads to problems with containing very hot material. ✓

$^{2}_{1}\text{H} + ^{2}_{1}\text{H} \rightarrow ^{3}_{2}\text{He} + ^{1}_{0}\text{n}$ ✓

Change in mass $\Delta m = 2 \times 2.01410 \text{ u} - 3.01603 \text{ u} - 1.00867 \text{ u}$

$= 0.00350 \text{ u}$ ✓

$= 0.00350 \times 1.67 \times 10^{-27} \text{ kg} = 5.85 \times 10^{-30} \text{ kg}$ ✓

$\Delta E = c^2 \Delta m = (3.0 \times 10^{8} \text{ m s}^{-1})^2 \times 5.85 \times 10^{-30} \text{ kg}$ ✓

$= 5.26 \times 10^{-13} \text{ J}$ ✓

This is the energy from the fusion of two nuclei, so energy from 1.0 kg = $5.26 \times 10^{-13} \text{ J} \times 1.5 \times 10^{26} = 7.89 \times 10^{13} \text{ J}$ ✓✓

Any sensible statement, e.g. ✓

- fuel for fusion can be obtained from water
- fusion reactors produce less radioactive waste than fission reactors.

4 *This is an example of an unstructured question involving estimation. If such a question gives a hint of how to start, it is sensible to take the advice!*

The examiners' mark scheme contains a range of points that would score marks. In the first part of this question you could score full marks with any eight of the following points.

Estimating the volume

$V = 13 \times 10^3 \text{ m} \times 13 \times 10^3 \text{ m} \times 33 \times 10^3 \text{ m}$ minus a bit for rounded edges ✓

$= 5 \times 10^{12} \text{ m}^3$ ✓

Take care converting from km to m. It is easiest to convert the length and width to m before multiplying.

Any estimate leading to a few times 10^{12} m would gain marks

Mass $m = \rho V = 2700 \text{ kg m}^{-3} \times 5 \times 10^{12} \text{ m}^3$ ✓

$= 1.4 \times 10^{16}$ kg ✓

For circular orbit of radius r, centripetal force $F = mv^2/r$ ✓

which is provided by gravitational force $F = GMm/r^2$ ✓

Hence $\frac{mv^2}{r} = \frac{GMm}{r^2}$ and so $v^2 = \frac{GM}{r}$ ✓

$$v^2 = \frac{6.67 \times 10^{-11} \text{ N m}^2 \text{ kg}^{-2} \times 1.4 \times 10^{16} \text{ kg}}{35 \times 10^3 \text{ m}}$$ ✓

so $v = 5 \text{ m s}^{-1}$ ✓

Your answer will depend on your estimate of mass.

Any one sensible suggestion would score a mark, e.g. ✓

- a large force is required to act in exactly the right direction
- Eros has a weak gravitational field so even a slightly too high speed would drive the craft away from its orbit.

5 Half-life is the average time for half the nuclei to decay ✓

$dN/dt = \lambda N = N \ln(2)/t_{1/2}$ ✓

$t_{1/2} = 1.3 \times 10^9 \text{ years} = 1.3 \times 10^9 \times 365 \times 24 \times 60 \times 60 \text{ s}$ ✓

$= 4.10 \times 10^{16}$ s

so $dN/dt = 3.0 \times 10^{15} \times \ln(2)/4.10 \times 10^{16} \text{ s} = 0.05$ Bq ✓

To get activity in Bq you must express $t_{1/2}$ in seconds.

One mark for any sensible comment, e.g. ✓

The activity is low so it would be difficult to detect it against the natural background radiation.

Two marks for any two sensible suggestions, e.g. ✓✓

- equipment would need to be shielded from background
- measurement would need to take place over a long period.

Assuming ✓

- no argon has escaped from the sample
- there was no argon initially in the sample.

Total number of potassium + argon nuclei = 2.4×10^{16} ✓

Originally all these nuclei were potassium. Now only one-eighth remain. The number has halved, halved then halved again. So the time elapsed must be $3 \times t_{1/2}$. ✓

So age = $3 \times 1.3 \times 10^9$ years = 3.9×10^9 years ✓

Alternatively you could use $N = N_0 e^{-\lambda t}$ and $\ln (N_0/N) = \lambda t$ to reach the same answer.

6 Nano means 10^{-9} ✓

$pV = nRT$ ✓

$$V = \frac{4\pi r^3}{3} = \frac{4\pi \times (0.5 \times 10\ \text{m})^3}{3} = 5.2 \times 10^{-22}\ \text{m}^3$$ ✓

$$n = \frac{pV}{RT}$$

$$= \frac{1.0 \times 10^8\ \text{Pa} \times 5.2 \times 10^{-22}\ \text{m}^3}{8.31\ \text{J mol}^{-1}\ \text{K}^{-1} \times 310\ \text{K}}$$ ✓

$= 2.0 \times 10^{-17}$ moles ✓

When pressure = $p_1 = 1.0 \times 10^5$ Pa,

volume required per minute = $V_1 = 2.5 \times 10^{-6}\ \text{m}^3$

If pressure = $p_2 = 1.0 \times 10^8$ Pa,

then $V_2 = p_1V_1/p_2 = 2.5 \times 10^{-9}\ \text{m}^3$ ✓

Volume per hour = $60 \times 2.5 \times 10^{-9}\ \text{m}^3 = 1.5 \times 10^{-7}\ \text{m}^3$ ✓

$1\ \text{m}^3 = 10^6\ \text{cm}^3$ so $1.5 \times 10^{-7}\ \text{m}^3 = 1.5 \times 10^{-1}\ \text{cm}^3 = 0.15\ \text{cm}^3$ ✓

Remember to convert to cm^3 as instructed in the question.

7 Redshift (accept Doppler shift) ✓

It is receding/moving away ✓

λ read from graph (allow plus/minus 10 nm) ✓

calculation of $\Delta \lambda/\lambda$ ✓

calculation of v ✓

It does not matter which line you choose. The table on the next page shows the values relating to all four lines.

emit. λ/nm	obs. λ/nm	$\Delta\lambda$/nm	$z = \Delta\lambda/\lambda$	$v = c\,\Delta\lambda/\lambda = cz$/m s^{-1}
410	475	65	0.159	4.8×10^7
434	505	71	0.164	4.9×10^7
486	560	74	0.152	4.6×10^7
656	760	104	0.159	4.8×10^7

$$d = \frac{v}{H_0}$$

$v = 4.8 \times 10^7$ m s^{-1} $= 4.8 \times 10^4$ km s^{-1} ✓

$$d = \frac{4.8 \times 10^4 \text{ km s}^{-1}}{75 \text{ km s}^{-1} \text{ Mpc}^{-1}} = 640 \text{ Mpc}$$ ✓

A galaxy at twice the distance would have twice the redshift (e.g. the shortest-wavelength line would be shifted by 130 nm to 540 nm). ✓

The flux would be weaker (it would be one-quarter that of 3C273). ✓

Unit PSA6

1 a Diffraction of electrons and/or X-rays. ✓

A clear structured pattern indicates order in the crystal. ✓

▶▶ *DIG Section 2.1, SUR Section 2.3*

b $v = \frac{c}{n}$ ✓

$$= \frac{3.00 \times 10^8 \text{ m s}^{-1}}{1.65} = 1.82 \times 10^8 \text{ m s}^{-1}$$ ✓

▶▶ *MUS Section 2.3, EAT Section 4.1*

c The electric field vibrations are confined to one plane. ✓

▶▶ *EAT Section 4.2*

d Use a polariser to produce polarised light. ✓

Send this light through a second polariser, and rotate until emerging light is either minimum (or maximum) brightness. ✓

Insert liquid sample between the two polarisers. ✓

If one of the polarisers needs to be rotated to get min (or max) brightness, then the plane of polarisation has been rotated by the liquid. ✓

Just saying 'use a polarimeter' would score one mark at the most. For four marks, examiners are looking for more detail.

▶▶ *EAT Section 4.2*

e $\sin 90° = 1$ so $d = \frac{1.05 \text{ mm}}{\Delta n}$ ✓

$$d = \frac{1.05 \text{ mm}}{0.15} = 7.00 \text{ mm}$$ ✓

f Assuming this is a uniform field ✓

$$E = \frac{V}{d}$$ ✓

$$= \frac{1.5 \text{ V}}{1.05 \times 10^{-6} \text{ m}} = 2.1 \times 10^{5} \text{ V m}^{-1}$$ ✓

▶▶ *MDM Section 5.2*

g Time for complete scan = $1/f = 1/25$ Hz ✓

$= 40 \times 10^{-3}$ s = 40 ms ✓

▶▶ *MUS Section 1.2*

Time for display to react/switch on is 10 ms ✓

so only four parts of the screen can react before the start of the next scan. ✓

h Marks for any *four* of these points.

Multiplexing means more than one signal ✓

using the same transmission medium/along the same channel/same fibre/same path. ✓

Frequency division multiplexing:

- signals of different frequencies ✓
- delivering information simultaneously. ✓

Time division multiplexing:

- signals sampled ✓
- and sent in succession along the same path. ✓

Transmitter and receiver need to be synchronised in order to decode signals correctly. ✓

▶▶ *MDM Section 2.4*

i Marks for any *two* of the following: ✓✓

- At higher temperature, the molecules become disordered (normal liquid) or a gas.
- At lower temperature the molecules become more ordered/less mobile (normal solid).
- At lower temperature viscosity increases so T_{on} becomes too large.

▶▶ *MDM Section 5.2, STA Section 3.3, EAT Section 2.2*

2 a There are two possible approaches, either of which can gain full marks.

Either Write the equation as

$$T_{on} = (k\eta d^2) \times \frac{1}{V^2}$$ ✓

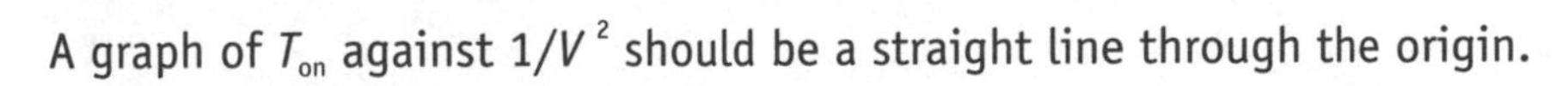

A graph of T_{on} against $1/V^2$ should be a straight line through the origin.

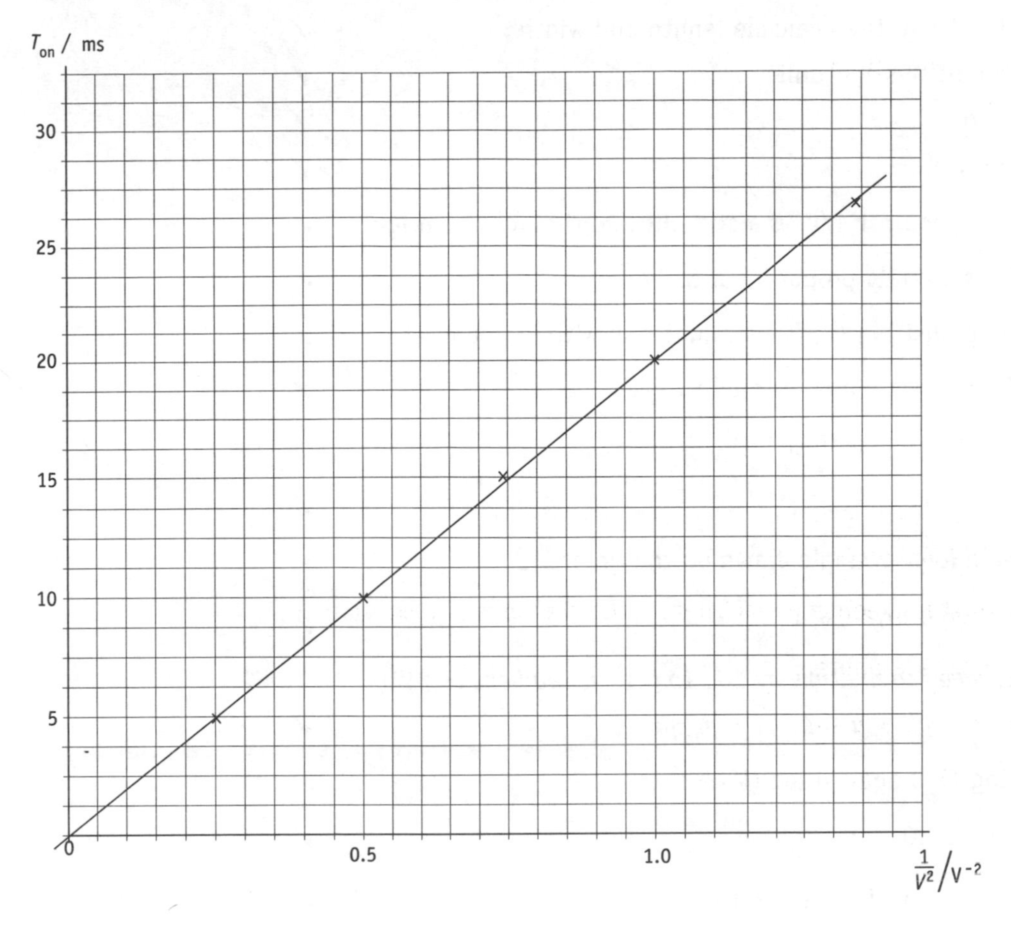

✓

List values of $1/V^2$ ✓

	turn-on time T_{on}/ms	voltage V/V	$(1/V^2)$/V^{-2}
	5	2.01	0.248
	10	1.42	0.496
	15	1.16	0.743
	20	1.00	1.00
	27	0.86	1.35

Or Take logs of both sides of the equation to get

$$\log(T_{on}) = \log(k\eta d^2) - 2\log(V)$$ ✓

A graph of log (T) against log (V) should be a straight line with gradient −2. ✓

List values of log (T_{on}) and log (V) ✓

log (T_{on}/s)	turn-on time T_{on}/ms	voltage V/V	log (V/V)
0.70	5	2.01	0.303
1.00	10	1.42	0.152
1.18	15	1.16	0.064
1.30	20	1.00	0.000
1.43	27	0.86	−0.066

Marks for graph plotting:

- sensible scales, at least half the available length and width ✓
- axes labelled with quantities and units ✓
- points plotted correctly ✓
- 'best' straight line drawn. ✓

b *Either* A graph of T_{on} against $1/V^2$ *is* a straight line through the origin ✓

which confirms that T_{on} *is* directly proportional to $1/V^2$. ✓

Or A graph of log (T) against log (V) *is* a straight line with gradient −2. ✓

which confirms that T_{on} *is* directly proportional to $1/V^2$. ✓

c *Either* Use the graph of T_{on} against $1/V^2$.

Gradient = $k\eta d^2$ ✓

Find the gradient using a *large* triangle drawn on the graph.

Gradient m = 20 ms V^2 = 20×10^{-3} s V^2 ✓

T_{on} is given in ms so you need to multiply by 10^{-3} to get the gradient in s V^2.

$$k\eta d^2 = m \quad \text{so} \quad k = m/\eta d^2$$ ✓

Or Use the graph of log (T_{on}) against log (V)

Intercept $c = \log(k\eta d^2) = 1.30$ ✓

$k\eta d^2$ = antilog (1.30) = 19.9 ms V^2 = 19.9×10^{-3} s V^2 ✓

From the original equation, $k\eta d^2$ has the same units as $V^2\,T_{on}$, i.e. ms V^2.

$$k = \frac{\text{antilog}\,(c)}{\eta d^2}$$ ✓

In either case:

$$k = \frac{20 \times 10^{-3}\ \text{s V}^2}{0.072\ \text{Pa s} \times (6.0 \times 10^{-6}\ \text{m})^2} = 7.7 \times 10^9\ \text{V}^2\ \text{N}^{-1}$$ ✓

As 1 Pa = 1 N m^{-2}, the units underneath the calculation are N $m^{-2} \times m^2$ which reduce to N.

Remember to convert d from mm into m, and then to square it.

d As T_{on} is proportional to d^2,

either plot T_{on} against d^2 (y-axis is T_{on} and x-axis d^2)

or plot log (T_{on}) against log (d) (y-axis is log (T_{on}) and x-axis log (d))

to get a straight line. ✓

Use the same viscosity/same liquid crystal at the same temperature (h). ✓

Use the same voltage (V). ✓

Index